Metamorphic Processes

Other Geology Titles:

Allen: PHYSICAL PROCESSES OF SEDIMENTATION
Hatch, Rastall and Greensmith: PETROLOGY OF THE SEDIMENTARY ROCKS
Hatch, Wells and Wells: PETROLOGY OF THE IGNEOUS ROCKS
Miyashiro: METAMORPHISM AND METAMORPHIC BELTS
Read: RUTLEY'S ELEMENTS OF MINERALOGY

Metamorphic Processes

Reactions and Microstructure Development

R. H. VERNON

Macquarie University, Sydney, Australia

A HALSTED PRESS BOOK

JOHN WILEY & SONS
New York

Published in the U.S.A.
by Halsted Press, a Division
of John Wiley & Sons, Inc.
New York

First published in 1975

ISBN: 0 470 90655-3

Library of Congress Cataloging in Publication Data
Vernon, Ron H.
   Metamorphic processes.
   'A Halsted Press book.'
   Includes bibliographies and index.
   1. ✓ Metamorphism (Geology)   I. Title.
QE475.A2V47   1975   552'.4   75-9139
ISBN 0-470-90655-3

Set in 10 point Times Roman type
by Linocomp Ltd., Marcham, Oxon.

Printed and bound in Great Britain by
REDWOOD BURN LIMITED
Trowbridge & Esher

To Judy

Preface

This book is for senior undergraduate or postgraduate students who want an insight into some modern approaches to metamorphic petrology. Its aims are to explain, in reasonably simple, informal terms, the *processes* underlying (i) metamorphic reactions and (ii) the production of microstructures in metamorphic rocks, these currently being the things that interest me most, geologically. The first aim requires discussion of equilibrium factors, reaction kinetics and reaction mechanisms, emphasising both the complexity of *realistic* reactions and the need to *combine* the chemical and microstructural approaches to them. The second aim requires discussion of deformation, recovery, recrystallisation and grain growth processes, with emphasis on experiments on silicate minerals. The book concludes with a general attempt to relate chemical and physical processes in metamorphism, although it will be clear from reading earlier chapters (especially Chapter 4) that the two aspects can rarely be separated completely in detailed metamorphic studies.

Petrological and experimental investigations of metamorphic reactions and microstructural development are advancing so rapidly these days that students are faced with an ever-increasing volume of information and a relatively rapid obsolescence of data. So, in this book I do not try to be comprehensive, or to present much so-called 'factual' information. Instead, I deal more with basic principles, in the hope that these will guide the student in his or her encounters with the details of specific metamorphic problems. I have selected for more detailed discussion certain reactions and deformation-recrystallisation studies on certain minerals, merely as examples of the approaches being used. The phase diagrams used are minimal, and mainly are rationalisations of calculated and experimental data that seem reasonable to me at this time; once again, they are used only as examples of an approach.

The references will enable a keen student to move into the recent literature, where he or she will find the evidence of human endeavour and conflict that characterise modern metamorphic petrology, just as they do all aspects of science, but which cannot be conveyed adequately in a concise book of this kind. The main references are listed for convenience at the end of each chapter, and are referred to in the text and figure captions by superscript numbers. Additional references of more specific, rather than general, interest appear in some figure captions. Abbreviations used through the book are explained on p. 15.

I wish to thank Ray Binns, Ian Duncan, Dick Flood, Vic Wall, Paul Williams and Chris Wilson for critically reading various parts of the typescript. Special thanks are due to Eric Essene and Vic Wall for permission to use some of their unpublished phase diagrams (Figs 2.7, 2.14, 4.4, 4.5 and 4.12), to Dick Yund for permission to use figure 3.8, to Mike Etheridge for permission to use figure 7.4 and to Ian Duncan, whose critical mind not only improved the book, but makes me wonder how he was able to tolerate the lectures on which the first draft was based. I also sincerely thank Alison Coates, Tina Terpstra, Nanette Madjlessi-Leenen, Sherry Knight, Dean Oliver, Rod Bashford and Dick Flood for much appreciated assistance in the preparation and organisation of typescript and drawings. I am particularly grateful to Alison Coates for an incredible amount of typing, and for ensuring that my spelling and grammar maintained a semblance of respectability. The book was begun at Macquarie University and finished at the University of Leiden (through the courtesy of Professor H. J. Zwart), and I appreciate the technical help provided by these institutions.

RON VERNON

School of Earth Sciences
Macquarie University
Sydney, Australia

April 1974

Contents

Abbreviations

G	Gibbs free energy
Δ	change of a quantity involved in a reaction
T	temperature (any scale in general context; absolute scale in thermodynamic equations)
P	confining pressure
V	volume (in general use)
V	vapour (when used in chemical reactions)
S	entropy
X_i	mole fraction of component i in a solution
a_i	activity of component i
f_i	fugacity of component i
n_i	number of moles of component i
μ	chemical potential
μm	micron ($=10^{-6}$ m)
kb	kilobar ($=10^8$ pascals)
kc	kilocalorie
R	universal gas constant
D	diffusion coefficient
Q	activation energy
ln	natural logarithm
K	equilibrium constant
P	number of phases
C	number of components
F	number of degrees of freedom
b	Burgers vector of a dislocation
γ	interfacial free energy
$\bar{\sigma}$	mean stress
$\sigma_1, \sigma_2, \sigma_3$	principal stress axes
$\lambda_1, \lambda_2, \lambda_3$	principal axes of finite strain
$\dot{\epsilon}$	strain rate

P, C, F : in the Phase Rule

Chapter 1

Background Discussion

Introduction

I assume you have encountered metamorphic minerals and rocks in elementary geology courses and subsequent petrology courses, and that you are aware of leading text-books on metamorphism, treated mainly from the chemical, mineralogical or field viewpoints.[9, 12, 16] In this book I want to concentrate on the details of metamorphic reactions and micro-structures, especially those aspects that require a combination of chemical and physical processes for their fuller understanding. In this chapter I will discuss a few aspects of metamorphism that are a necessary background to the rest of the book.

Scope of Metamorphism

Rock alteration in the solid state can occur in a wide range of geological environments. Some types of rock alteration are traditionally separated from metamorphism. For example, weathering can be so excluded, because it occurs at atmospheric conditions. However, other alteration processes grade into metamorphism as it is defined by some people. For example, what is generally called 'diagenesis' of sediments (cementation and altera-tion of sedimentary minerals in response to burial) grades into incipient, so-called 'burial metamorphism'. However, apart from very early dia-genetic changes occurring at atmospheric conditions, most diagenetic reactions appear to me to warrant the term 'metamorphic', since the rocks remain predominantly solid and the P-T conditions are crustal, not atmospheric. So I regard the greater part of 'diagenesis' as *burial meta-morphism*, because I find it impossible to draw an effective line between the two.

Terms commonly used to describe specific metamorphic environments include 'burial metamorphism', 'wall-rock alteration' (around hydro-thermal orebodies), 'geothermal alteration' (in sub-volcanic geothermal areas) and 'deuteric alteration' (in largely or entirely crystallised igneous intrusions). These terms have value inasmuch as they indicate geological

17

environments in which metamorphism can take place. But they have no value when we are considering the conditions of formation of metamorphic mineral assemblages. The minerals react to temperatures, pressures and the effects of changing activities of mobile components, irrespective of how these constraints are imposed on them. So they react to the same set of imposed conditions in the same way, regardless of the geological environment responsible. For example, identical assemblages could be formed in (*a*) deuteric alteration in an igneous intrusion, (*b*) wall-rock alteration about a hydrothermal orebody, (*c*) lowest grade contact altera-tion around an igneous intrusion, (*d*) burial metamorphism of sedimentary/volcanic sequences and (*e*) alteration, both local and regional, in geothermal areas.

Metamorphism is generally divided into two main classes: *contact (thermal) metamorphism*, occurring around igneous intrusions; and *regional metamorphism*, in which metamorphic rocks are produced on a regional scale. However, the same minerals can be produced in both these types of metamorphism, so a recent tendency is to classify metamorphism into three *broad pressure-classes* grading into each other (Fig. 1.1) and transcending the contact/regional classification, although of course these terms still have geological value.[12]

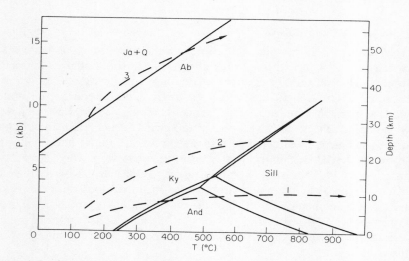

Fig. 1.1   Possible P-T gradients ('trajectories') for low-pressure (1), inter-mediate-pressure (2) and high-pressure (3) metamorphism superimposed on the Al_2SiO_5 phase diagram of Fig. 4.4 and the albite $\rightleftharpoons$ jadeite+quartz reaction curve (cf. Fig. 4.18), upon which the major pressure-classes of metamorphism are based. The scheme is based on Miyashiro's concept of metamorphic 'facies series'.

These pressure-classes are based on two sets of solid-solid equilibria, namely: (i) the Al_2SiO_5 phase relationships, such that metapelitic rocks containing apparently stable andalusite are called 'low-pressure' types,

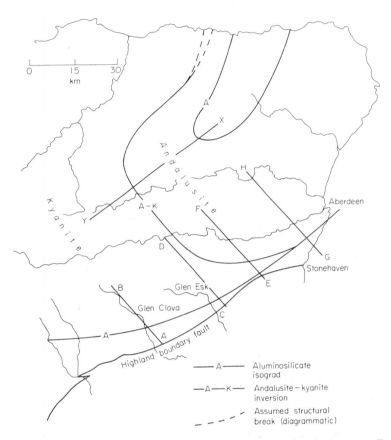

Fig. 1.2 Map showing major aluminosilicate isograds in the eastern Dalradian of Scotland. Traverses AB, CD, etc. correspond to those of Fig. 1.3. The 'aluminosilicate isograd' (A) is an andalusite isograd in the north and partly a kyanite, partly an andalusite isograd in the south.
After Porteous, p. 30 (Scottish Academic Press Ltd.).[11]

and metapelitic rocks containing stable kyanite are 'intermediate-pressure' types; and (ii) the albite $\rightleftharpoons$ jadeite + quartz transition, such that rocks containing jadeite + quartz are called 'high-pressure' types (Fig. 1.1).

A well-known example of apparently synchronous low-pressure and intermediate-pressure metamorphism occurs in the eastern Dalradian metamorphic region of the Scottish Highlands (Figs 1.2, 1.3), where an

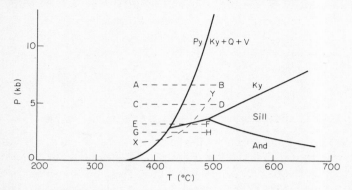

Fig. 1.3 Some aluminosilicate phase equilbria (from Figs 4.4 and 4.5), with traverses corresponding to those of Fig. 1.2.

The pyrophyllite $\rightleftharpoons$ kyanite + quartz + vapour curve is used as the 'aluminosilicate isograd' of Fig. 1.2, being the lower stability limit of kyanite + quartz (for $P_{H_2O} = P_{total}$), but in the Dalradian rocks the 'aluminosilicate isograd' may lie at higher (and variable) temperatures, depending on the actual reactions involved. Modified after Porteus, p. 35.[11]

area dominated by andalusite ('Buchan'-type or low-pressure metamorphism) occurs adjacent to a kyanite area ('Barrovian'-type or intermediate-pressure metamorphism). The two areas are inferred to be separated by an 'andalusite-kyanite inversion' line.[11] This also appears to be an example of the development of low-pressure metamorphism on a

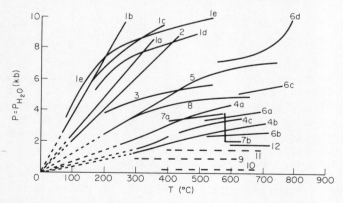

Fig. 1.4 Estimated P-T gradients in different areas, suggesting that each area may have a unique P-T gradient. After Turner, p. 359.[12]

The numbers refer to specific areas (see Turner for details if required).

regional scale, although the situation is complicated by the presence of many igneous intrusions.

Detailed study of the minerals and inferred reactions in many areas, coupled with reference to experimental information (Chapters 2, 4), has facilitated inference of a P-T gradient for each area. The great variety of possible gradients is well shown by figure 1.4. In fact, every area may have a unique P-T gradient.[12]

Temperatures and Pressures of Metamorphism

Experiments on metamorphic minerals and measurements in geothermal wells have shown that the lower temperatures of crustal metamorphism are around < 100–300°C, and that the upper temperatures are commonly around 700–750°C (locally as high as 900–1 000°C; Fig. 1.5).

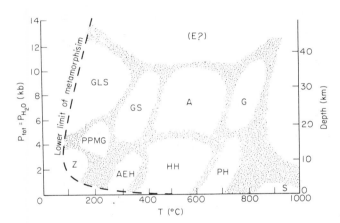

Fig. 1.5   Most commonly accepted metamorphic facies approximately plotted on a P-T diagram for $P_{total} = P_{H2O}$, showing gradational zones between them. Modified after Turner, p. 366.[12] Abbreviations representing the facies are:

AEH	albite-epidote hornfels
E	eclogite
G	granulite
GLS	glaucophane-lawsonite schist
GS	greenschist
HH	hornblende hornfels
PH	pyroxene hornfels
PPMG	prehnite-pumpellyite metagreywacke
S	sanidinite
Z	zeolite

Confining pressures (load pressures, lithostatic pressures) in crustal metamorphism, caused by the load of overlying rock, range from about 0 to 15 kilobars (1 bar = 10^5 pascals = 0·987 atmosphere), increasing by about 1 kb for each 3 km of depth in a column of rock of average sialic crustal density (2·7). This confining pressure (or hydrostatic stress) acts equally in all directions on a relatively large aggregate of grains, but local constraints may cause stress differences on the scale of individual grains.

In addition, a non-uniform component of pressure (differential stress) may be present, causing the rocks to deform permanently by ductile flow, producing characteristic structures such as foliations and folds. These structures are typical of regional metamorphism in orogenic zones. One of the aims of structural geology is to interpret these distortions (changes of shape or *strains*) of the rocks. Metamorphic petrology is concerned with the relationships of these structures to mineral growth and deformation, and with the possible effect of differential stress on chemical reactions.

Movement of Material in Metamorphism

The change from an unstable assemblage of minerals to a stable one involves (i) release of atoms from their positions in the lattices of the unstable mineral grains, (ii) formation of nuclei of the stable minerals, (iii) movement of atoms to the nuclei, (iv) movement away from the reaction sites of atoms of unwanted by-products. Most authorities on metamorphism state that a *fluid phase* is always present during metamorphism (although the rocks remain solid), and this is probably true of porous sedimentary rocks. Such a fluid phase would facilitate transfer of material in metamorphic reactions. Many metamorphic reactions involve the production of water or carbon dioxide, which must be removed from the rock somehow, in order for the reactions to proceed without changing the temperature. Some other reactions require water to be introduced to the reaction site. So movement of 'volatile' or other ionic components takes place in metamorphism, at least on the scale of several grains (Chapters 2, 4), although the exact nature and physical state of these components in many metamorphic environments are arguable (Chapter 2).

If it is believed that a particular body of rock has not changed its overall chemical composition (apart from its water content, and ignoring transfers on the scale of a few grains) during metamorphism, the metamorphism is said to be 'isochemical'. Most metamorphism generally is considered to be isochemical, because most metamorphic rocks have chemical compositions broadly similar to common sedimentary or igneous rocks (except for H_2O and CO_2). If the bulk chemical composition of the rock has changed, the metamorphic is said to be 'allochemical' or 'metasomatic'. Metasomatism produces rocks of extreme or unusual chemical composition, but it is often difficult to prove that metasomatism has taken place over large distances. Diffusion of elements is common over small distances (a few centimetres or so) producing metamorphic layering or metasomatic zones between minerals or rocks of different composition that reacted with each other during metamorphism (Chapters 4, 8).

Mineral Assemblages ('Parageneses')

An important aim in metamorphic studies is to determine which minerals in a rock grew at the same time, i.e. were in stable, or metastable, co-existence during the metamorphic event being considered. Such a group of synchronous coexisting minerals sometimes is called a 'paragenesis'.[16] The recognition of such assemblages can be difficult, especially in some polymetamorphic rocks.

Microstructures of Metamorphic Rocks

The shapes and arrangement of grains in a metamorphic rock are its *microstructure* or '*texture*'. Most people use the word 'texture', because of tradition and because its meaning generally is not ambiguous. However, I prefer (without getting upset about it) the term 'microstructure', as used in physical metallurgy, because (i) 'texture' means 'preferred orientation' in synthetic materials, especially metals; (ii) in the modern approach to materials science, all crystalline materials (including rocks) tend to be grouped together in terms of basic processes; and (iii) 'microstructure' fits in well with the two other structural scale-terms in common use, namely 'mesostructure' and 'macrostructure'.

In sedimentary rocks the microstructure (texture) is generally dominated by the occurrence of fragments plus cementing material, and in igneous rocks the microstructure is dominated by a tendency towards an order of crystallisation. However, in metamorphic rocks the grains all grow at about the same time, so that the microstructure is dominated by *solid-state grain adjustments* (Chapters 5, 6), and, in many instances, by deformation as well (Chapters 5, 6 and 7).

Though some metamorphic microstructures look superficially similar to some igneous microstructures, they have an altogether different significance. For example, *porphyroblastic* structure in metamorphic rocks resembles porphyritic structure in igneous rocks, but may have little to do with relative cooling rates. Instead, the mineral growing as porphyroblasts (cf. igneous phenocrysts) may have formed only a few nuclei compared with all the other minerals present in the rock. Similarly, the metamorphic term *xenoblastic* means (as does the igneous term xenomorphic) that a grain has no crystal faces, but, whereas igneous minerals are xenomorphic because they have to fill irregularly shaped spaces between pre-existing crystals, metamorphic minerals are xenoblastic because of mutual interference of growing grains. An aggregate of xeno-blastic grains is said to be *granoblastic*, especially if the grains are pre-dominantly equant and polygonal in shape (Chapter 5).

In other words, the shapes of grains in metamorphic rocks depend on the ways in which the grains adjust to each other's presence in the solid state. This adjustment takes place at the grain boundaries, as discussed in detail later (Chapter 5).

Preferred Orientation in Metamorphic Rocks

In rocks formed or deformed in a 'stress field' (an environment in which the stresses acting on a small volume of rock are unequal), a *preferred orientation* of the metamorphic mineral grains may be produced (Chapters 6, 7). This is typical of regional metamorphism.

Granoblastic aggregates can show a preferred orientation of their atomic lattices without any elongation of the grain shapes. Detailed optical or X-ray measurements are needed to detect this lattice preferred orienta-tion. However, most preferred orientations involve parallel alignment of elongate grains (shape or dimensional preferred orientation), and are readily detected. In many rocks this parallel alignment is strong enough to see in hand specimen, either as a *foliation* (planar or sheet-like structure) or *lineation* (parallel alignment of prismatic minerals or rod-like aggregates of minerals).[13] Commonly a lineation lies in the plane of a foliation in the same rock. Foliations and lineations generally can be related to fold structures as discussed in structural geology textbooks.[5, 13]

Metamorphic 'Grade'

In many metamorphic areas, it has been found that the most deformed and/or recrystallised rocks contain mineral assemblages that differ from those in less strongly altered rocks in other parts of the areas concerned. This suggests that the intensity of metamorphism ('metamorphic grade') may vary from one part of an area to another. For example, in some areas, incipiently metamorphosed (low-'grade') rocks contain minerals which are stable at relatively low temperatures, as experience with igneous rocks and laboratory experiments tells us. Many of these minerals (e.g. chlorite, muscovite, talc) are hydrated, and many reactions that take place with increasing metamorphic grade (in a sequence of *progressive metamorphism* through a volume of rock undergoing metamorphism) involve dehydration of these hydrous minerals to give mineral assemblages containing less combined water. High-'grade' rocks in these areas contain minerals stable at high temperatures. A sequence of *metamorphic zones* (zones of different minerals marking differences in grade) may be mappable across an area of regional metamorphic rocks or in a contact metamorphicv aureole around an igneous intrusion (the lowest-grade zones occurring furthest away from the contact).

This common use of the term 'grade' correlates it mainly with temperature. However, temperature is not the only variable controlling metamorphic assemblages, other factors being pressure and the activities of mobile components (Chapter 2). For example, 'grade' variations in high-pressure metamorphic areas would be related much more to progressive changes in confining pressure than to temperature. So, though 'grade' is in general use, it is often vaguely defined, and, where used in a particular area, its exact nature should be clearly specified.

Lines on a map separating metamorphic zones are called *isograds* (lines of equal 'grade'). Traditionally isograds are based on the first appearance of an *index mineral*, after which the zone is named; for example, the following sequence of progressive metamorphic zones has been used for over fifty years in the southeast Scottish Highlands: chlorite, biotite, garnet, staurolite, kyanite and sillimanite zones. The distribution of Al_2SiO_5 isograds in eastern Scotland is shown in figure 1.2. However, more detailed work in some areas is showing that *assemblages* of minerals may be more useful as isograd markers than single minerals (Chapters 2, 4).

The mineral assemblage developed in a metamorphic rock depends on two main factors: (i) the chemical composition of the rock and its fluid phase (if present); (ii) the P-T conditions. Note that some rocks are more reactive, and hence are better 'grade' indicators, than others.

Metamorphic Facies

A metamorphic facies is a collection of metamorphic mineral assemblages, repeatedly associated in space and time, such that a constant relationship exists between mineral assemblage and bulk chemical composition of the rock.[12] Note that:

(i) A metamorphic facies does not refer to a single rock-type, but embraces a number of rock-types (or mineral assemblages), all formed under the same broad conditions of temperature, confining pressure and other variables (e.g. composition of fluids present, if any).

(ii) All rocks of the same bulk chemical composition have the same mineral assemblage if they belong to the same metamorphic facies.

(iii) The mineral assemblages concerned must be stable or consistently metastable wherever found throughout the world.

(iv) The metamorphic facies scheme is useful as a broad genetic classification of metamorphic rocks, in terms of the major variables: confining pressure and temperature; it is especially useful for regional or reconnaissance studies in metamorphic regions, but is too broad for many detailed metamorphic studies.

(v) Various facies schemas are in use, but probably the most commonly accepted scheme is that of figure 1.5. Some controversy surrounds the 'eclogite facies' (hence the question mark), but to me it seems reasonable to follow Turner and include it as representing a large high P-T regime in the lower crust and upper mantle.[12] Note, however, that not all 'eclogites' (i.e. garnet-clinopyroxene rocks) are formed in this regime.

(vi) Owing to complexity of reactions and the general likelihood of boundary reactions occurring over P-T intervals, rather than on sharp P-T lines (Chapter 2), the boundaries between metamorphic facies should be regarded as being gradational (Fig. 1.5).

(vii) Mineral assemblages diagnostic of each facies are given in Table 1.1. Note that generally these occur in rocks of *mafic* composition, although some distinctions are based on minerals in *metapelitic* rocks.

(viii) Correct application of the facies classification cannot be made in areas devoid of rock compositions suitable for forming these diagnostic assemblages.

(ix) Table 1.1 shows how difficult it can be to distinguish between the facies of lower pressure and their intermediate-pressure equivalents (e.g. between the albite-epidote hornfels facies and the greenschist facies). This is especially so in the absence of detailed chemical data (which may assist in the distinction between granulite and pyroxene hornfels facies; for example, on the basis of pyroxene compositions), and such data are outside the range of interest of people to whom the facies classification is most useful.

(x) The standard facies schemes (e.g. Fig. 1.5) apply only to the situation where water is the only fluid component present, and for $P_{H2O} = P_{total}$. Recent experimental, theoretical and petrographic work has suggested that mineral assemblages of the zeolite, prehnite-pumpellyite metagreywacke and glaucophane-lawsonite schist facies are stable only if any fluid phase present contains low concentrations of CO_2.[2,3] So, if figure 1.5 were re-drawn for the situation in which CO_2 made up an appreciable part of the fluid phase (say a half or even less), these three facies would not appear at all, because their diagnostic assemblages would be unstable. Instead, their place would be taken by an extended greenschist facies. Under these conditions, progressive (prograde) metamorphism would take original sedimentary or igneous rocks straight into the greenschist facies, as in the present-day geothermal area of the Salton Sea, California, where CO_2 is abundant.[4, 10] Furthermore, the same diagram would show considerable expansion of the eclogite facies and would show shifts of the

Table 1.1: *Assemblages Diagnostic of Metamorphic Facies*

Facies	*Mineral Assemblage*
Zeolite	Zeolites, especially laumontite
Prehnite-pumpellyite metagreywacke	Prehnite + pumpellyite
Glaucophane-lawsonite schist	Glaucophane + lawsonite; jadeite + quartz; (aragonite)
Albite-epidote hornfels	Albite + epidote + tremolite / actinolite ($\pm$ chlorite $\pm$ calcite) in mafic rocks. Andalusite may occur at higher grades in metapelitic rocks; pyrophyllite
Green-schist	Ditto; stilpnomelane may be diagnostic; no andalusite; Pyrophyllite in metapelitic rocks
Hornblende hornfels	Hornblende + plagioclase ($> An_{20}$) in mafic rocks Andalusite in metapelitic rocks
Amphibolite	Hornblende + plagioclase ($> An_{20}$); almandine and epidote more common than in hornblende hornfels facies (mafic compositions) Kyanite in metapelitic rocks
Pyroxene hornfels	Orthopyroxene in mafic compositions
Granulite	Ditto; pyrope-almandine garnet more common than in pyroxene hornfels facies
Eclogite	Jadeite-rich clinopyroxene (omphacite) + pyrope-rich garnet in mafic compositions; pyrope-rich garnet in ultramafic compositions

greenschist-amphibolite and amphibolite-granulite boundaries (along with their low-pressure equivalents) towards lower temperatures, because of the effect of reduced activity of water on the relevant dehydration reactions (discussed in detail in Chapter 2). This means that a complete representation of metamorphic facies should involve relationships in P-T-X space (where X = mole fraction of CO_2 in an H_2O fluid phase*). Then we can

*The nature of the fluid phase and the activity of mobile components in metamorphism are discussed in more detail in Chapter 2.

consider P-T-X gradients, rather than P-T gradients, which may be essential for explaining relationships between lower grade facies. For example, the change from prehnite-pumpellyite metagreywacke facies to greenschist facies can occur simply by changing the fluid composition without changing P or T.[3]

(xi) From this it follows that local variations in fluid composition (especially the CO_2:H_2O ratio) could cause changes in mineral assemblages, owing to control of the fluid composition either by (*a*) reactions in local rock systems or (*b*) compositions of external fluid reservoirs near fluid-filled fractures etc. This situation could be interpreted as indicating more than one metamorphic facies in a small field area, outcrop, 'hand specimen' or normal-sized thin section. For example, orthopyroxene-rich pods occur rarely in some otherwise typically amphibolite facies gneisses in Antarctica.[14] Probably this is mainly due to a local reduction in water and possibly an increase in activity of silica, and not to a local P-T fluctuation, which would be a less reasonable interpretation. So, because the rest of the rocks in the region fit the amphibolite facies, it seems best to regard the region as belonging to the amphibolite facies in terms of broad P-T conditions.[14] The alternative is to abandon the facies scheme, which would be a pity for more general geological work, in which facies are used as broad P-T indicators. The situation would be more serious in areas where rocks of two facies are interlayered so intimately and repeatedly that neither facies is dominant over a region. Generally, however, a region can be assigned to a predominant facies, if the diagnostic rock-types are available.

(xii) Local increases in non-hydrostatic stress could cause stable or metastable growth of high-pressure phases in areas otherwise undergoing metamorphism in P-T conditions of the zeolite or greenschist facies.[1] But, once again, if this can be recognised as a local phenomenon only, the area as a whole can be ascribed to the relevant, predominant facies.

(xiii) Use of the facies scheme as a general or reconnaissance concept should not detract from more detailed, academic attempts to observe and interpret local assemblage variations. Moreover, as pointed out above, the facies scheme may be inappropriate for classifying assemblages in studies of this type, where the variations all occur at the same P-T conditions, unless extended into P-T-X space and used on a local scale.

Tectonic Setting of Metamorphism

Investigations into the tectonic setting of metamorphic events on a world scale have gained great impetus in recent years, owing especially to the development of the plate tectonics hypothesis. These are outside the scope of this book, which is concerned with fundamental meta-morphic processes on a much finer scale. Most geology courses discuss these topics nowadays, and the interested reader is referred to some leading publications on the evolution of the Archaean crust[15] and the distribution and relation to plate movements of more modern meta-morphic environments (including paired metamorphic belts and ocean-floor metamorphism).[6,7,8,9]

References

1 Brown, W. H., Fyfe, W. S. and Turner, F. J. (1962). Aragonite in Cali-fornia glaucophane schists, and the kinetics of the aragonite-calcite trans-formation. *J. Petrology*, **3**, 566–82.

2 Ernst, W. G. (1972). CO_2-poor composition of the fluid attending Fran-ciscan and Sanbagawa low-grade metamorphism. *Geochimica et Cos-mochimica Acta*, **36**, 497–504.

3 Glassley, W. (1974). A model for phase equilibria in the prehnite-pumpellyite facies. *Contribs. Mineralogy & Petrology*, **43**, 317–32.

4 Helgeson, H. C. (1968). Geologic and thermodynamic characteristics of the Salton Sea geothermal system. *Amer. J. Science*, **266**, 129–66.

5 Hobbs, B. E., Means, W. H. and Williams, P. F. (1975). *An Outline of Structural Geology*. New York: J. Wiley & Sons.

6 Miyashiro, A. (1961). Evolution of metamorphic belts. *J. Petrology*, **2**, 277–311.

7 Miyashiro, A. (1972). Pressure and temperature conditions and tectonic significance of regional and ocean-floor metamorphism. *Tectonophysics*, **13**, 141–59.

8 Miyashiro, A. (1973). Paired and unpaired metamorphic belts. *Tectono-physics*, **17**, 241–51.

9 Miyashiro, A. (1973). *Metamorphism and Metamorphic Belts*. London: George Allen & Unwin.

10 Muffler, L. J. P. and White, D. E. (1969). Active metamorphism of Upper Cenozoic sediments in the Salton Sea geothermal field and the Salton Trough, southeastern California. *Bull. Geol. Soc. America*, **80**, 157–82.

11 Porteus, A. (1973). Metamorphic index minerals in the eastern Dalradian. *Scottish J. Geology*, **9**, 29–43.

12 Turner, F. J. (1968). *Metamorphic Petrology. Mineralogical and Field Aspects*. New York: McGraw-Hill Book Co.

13 Turner, F. J. and Weiss, L. E. (1963). *Structural Analysis of Metamorphic Tectonites*. New York: McGraw-Hill Book Co.

14 Williams, P. F., Hobbs, B. E., Vernon, R. H. and Anderson, D. E. (1971). The structural and metamorphic geology of basement rocks in the McMurdo Sound area, Antarctica. *J. Geol. Soc. Australia*, **18**, 127–42.

15 Windley, B. F. and Bridgwater, D. (1971). The evolution of Archaean low- and high-grade terrains. *Special Pubs. Geol. Soc. Australia*, no. 3, 33–46.

16 Winkler, H. G. F. (1967). *Petrogenesis of Metamorphic Rocks*, 2nd Edn., Berlin-Heidelberg-New York: Springer-Verlag.

Chapter 2

Equilibrium Aspects of Metamorphic Reactions

Introduction

Examination of a metamorphic rock reveals (1) the mineral assemblage and (2) the distribution of elements between minerals, if detailed chemical analyses are carried out. Our problem is to infer from this information the metamorphic conditions that operated at the time of formation of the mineral assemblage. These conditions are mainly confining pressure (P), temperature (T) and mole fraction (reflecting the activity*) of mobile components (X). Ultimately, we would like to place our assemblage precisely in a P-T-X diagram (e.g. Fig. 2.17) for the particular bulk chemical composition of the portion of rock concerned (i.e. the chemical system).

Approaches to this problem are (1) field–laboratory relationships, which may give a general indication of relative metamorphic grade and suggest relevant metamorphic reactions; (2) theoretical calculation of proposed reaction curves based on thermodynamic properties of the minerals; and (3) experimental determination of the stability limits of the observed mineral assemblage or of inferred metamorphic reactions.

*Activity (a_i), chemical potential (μ_i) and fugacity (f_i) are all expressions for the behaviour of a component (i) in a solution (solid or fluid). I will use a_i for solid (ionic) components and f_i for gaseous (or supercritical) components. Concentration (expressed as the mole fraction of i, X_i, where X_i=no. of moles of i/total no. of moles in solution) is related to chemical potential (μ_i) in an *ideal* solution by: $\mu_i = \mu_i° + RT\ln X_i$, where $\mu_i°$ is a constant (the partial molar Gibbs free energy of pure i in a standard state). In a non-ideal solution, the relationship is: $\mu_i = \mu_i° + RT\ln a_i$, so that the activity is a number relating the *actual* and ideal behaviour of i in the solution. The relationship between a_i and X_i is given by: $\gamma_i = a_i/X_i$, where γ_i is called the activity coefficient. For gases, the behaviour of i is conveniently expressed in units of pressure, or, more correctly, fugacity (f_i), where f_i is related to activity by: $a_i = f_i/f_i°$, where $f_i°$ is the fugacity of i in a standard state. For an ideal solution, $f_i = a_i = X_i$. However, in real solutions, a_i (and hence f_i) may differ greatly from X_i.

Driving Force for Metamorphic Reactions

As for any chemical reaction at constant P and T, the driving force for metamorphic reactions is the Gibbs free energy change (ΔG).* For a spontaneous incremental change at constant T and P, ΔG must be either negative (for an irreversible reaction) or zero (at equilibrium), so that the total Gibbs free energy of the system (G) is a minimum at equilibrium.

Consider three chemically equivalent mineral assemblages in a metamorphic rock, namely $(A+B)$, $(C+D)$ and $(E+F)$, the following reactions being possible theoretically:

$$A+B \rightarrow C+D$$
$$A+B \rightarrow E+F$$
$$C+D \rightarrow E+F$$

If these are all spontaneous irreversible reactions for the relevant T and P (ΔG negative), then the thermodynamically stable assemblage must be $(E+F)$. Assemblages $(A+B)$ and $(C+D)$ are said to be *metastable* if they show no tendency to change to $(E+F)$. However, we can see that, if $(A+B)$ reacts to form $(C+D)$, the free energy of the system is lowered, so that this reaction could occur spontaneously. Nevertheless, the most stable state, namely $(E+F)$, has not been reached, and may never be if $(C+D)$ is sufficiently stable to persist under the prevailing conditions.

This brings up the question of *reaction kinetics,* to which we will return later (Chapter 3). Kinetic factors decide whether or not a reaction will take place. Thermodynamics can tell us only whether or not it can take place.

Types of Metamorphic Reactions

The following broad types of metamorphic reaction can be recognised,

---

*As defined in chemical[14] and geological[22,37,41] writings, the Gibbs free energy (G to most Europeans and Australians, F to most Americans) of a closed system (i.e. one which does not exchange matter with its surroundings) is the expression: $G=E+PV-TS=H-TS$, where E=internal energy, V= volume, T=absolute temperature, S=entropy, and H=enthalpy of reaction. It can be shown that $d(\Delta G)=(\Delta V)dP - (\Delta S)dT$, where Δ is the difference between reactants and products, so that, for a reaction at constant temperature ($dT=O$), the condition of equilibrium ($dG=O$) implies that the pressure must remain constant also ($dP=O$). So, the Gibbs energy function is an equilibrium indicator for reactions at constant pressure. A different free energy function must be used for reactions at constant volume (which is unlikely in metamorphic reactions, owing to the different molar volumes of reactants and products in most reactions; i.e. $dV \neq O$).

namely: (i) solid → solid + vapour reactions, and (ii) solid → solid reactions.

Solid → solid + vapour ('devolatilisation', 'devaporisation') reactions

Examples are:

$$Mg(OH)_2 \rightarrow MgO + H_2O \text{ ('dehydration')}$$
$$MgCO_3 \rightarrow MgO + CO_2 \text{ ('decarbonation')}$$

These reactions are characterised by (*a*) large positive ΔV for the forward reaction (because of the conversion of crystal-bound $(OH)^-$ or $(CO_3)^{2-}$ to a vapour phase occupying a much larger volume), (*b*) large positive ΔS (because of the change from relatively ordered $(OH)^-$ or $(CO_3)^{2-}$ in crystal lattices to highly disordered molecules in a vapour), and (*c*) large negative ΔG.

Dehydration reactions are common in the *prograde* (progressive) metamorphism of initially hydrous rocks (e.g. shales). However, if the initial rock is anhydrous (e.g. basalt), *hydration* reactions are involved in prograde metamorphism (producing minerals such as amphibole, chlorite and epidote). Dehydration may or may not follow this initial hydration, with further increase in metamorphic grade.

Eugster has distinguished between thermal (gas) and ionic equilibria.[18] For example, the *thermal stability* of muscovite in the presence of quartz is given by:

$$KAl_3Si_3O_{10}(OH)_2 + SiO_2 \rightleftharpoons KAlSi_3O_8 + Al_2SiO_5 + H_2O \qquad (1)$$
$$\text{muscovite} \qquad\quad \text{quartz \ K-feldspar \ sillimanite}$$

This reaction has been studied in the laboratory using pure H_2O fluid as the pressure medium (Fig. 2.5). In such experiments, the amount of solids dissolved in the fluid is negligibly small.

On the other hand, the *ionic stability* of muscovite in the presence of quartz is given by:

$$KAl_3Si_3O_{10}(OH)_2 + 6SiO_2 + 2K^+ \rightleftharpoons 3KAlSi_3O_8 + 2H^+ \qquad (2)$$

This reaction also has been studied experimentally, and proceeds at much lower temperatures than reaction (1).[30] Thermodynamic data for such reactions are given by Helgeson.[28, 29]

Some reactions involve both stable gases and ions, such as:

$$2KAl_3Si_3O_{10}(OH)_2 + 2H^+ + 3H_2O \rightleftharpoons 3Al_2Si_2O_5(OH)_4 + 2K^+ \qquad (3)$$
$$\text{muscovite} \qquad\qquad\qquad\qquad\qquad \text{kaolinite}$$

As discussed in Chapter 4, Carmichael has emphasised the likely importance of gas/ionic reactions in metamorphism.[8]

The available experimental data on reactions (1) and (2) can be combined to obtain equilibrium relationships among muscovite, quartz, K-feldspar and sillimanite (or andalusite or kyanite) in terms of P, T, f_{H_2O}, pH and a_{K^+}. This broadens considerably the range of variables controlling the stability of these solid phases and, therefore, gives us a more complete way of approaching metamorphic reactions involving a fluid phase.[18, 28]

By combining the equilibrium constants for reactions (1) and (2), Eugster derived equilibrium constants for two other reactions,[18] namely:

$$2KAl_3Si_3O_{10}(OH)_2 + 2H^+ \rightleftharpoons 3Al_2SiO_5 + 3SiO_2 + 2K^+ + 3H_2O \quad (4)$$
$$\text{muscovite} \qquad\qquad\qquad \text{sillimanite} \quad \text{quartz}$$

and

$$2KAlSi_3O_8 + 2H^+ \rightleftharpoons Al_2SiO_5 + 5SiO_2 + 2K^+ + H_2O \quad (5)$$
$$\text{K-feldspar} \qquad\qquad \text{sillimanite} \quad \text{quartz}$$

Since the equilibrium constants of both reactions are given by

$$\log K = \left(a^2_{K^+} \times f_{H_2O} \right) \Big/ \left(a^2_{H^+} \right)$$

we have all the information needed to evaluate relationships between the solids and a supercritical fluid of stable gases plus ions. If P, pH and f_{H_2O} are fixed, a $T - a_{K^+}$ diagram can be constructed (Fig. 2.1). K-feldspar occurs at high a_{K^+} for all temperatures, sillimanite occurs at low a_{K^+} and high T, and muscovite occurs at low a_{K^+} and low T.

From figure 2.1 we can see that, if the activity of K^+ ions in solution in equilibrium with K-feldspar is reduced (for example, by the nearby crystallisation of a K-rich phase), K-feldspar will tend to break down to muscovite by reaction (2), so that the muscovite field of figure 2.1(a) is entered. Exactly the same effect would be achieved by an increase in H^+ or decrease in $(OH)^-$ of the solution (for example, by the crystallisation of a hydrous silicate), which expands the muscovite field at the expense of K-feldspar (Fig. 2.1(c)). It may be impossible to distinguish the two processes and, in fact, they may both proceed together.

Similarly, figure 2.1 indicates that the muscovitisation of an Al_2SiO_5 polymorph, according to reaction (4), is favoured by increasing f_{H_2O}, pH, or a_{K^+}, and decreasing T. For example, as a_{K^+} in the fluid phase is decreased, muscovitisation occurs at lower temperatures. Even so, for $f_{H_2O} = 1\,000$ bars, the reaction would still occur at a temperature high

enough for at least partial alteration to occur. Therefore, to avoid alteration of Al_2SiO_5, either the a_{K^+} of the fluid phase must be extremely low
(well below 0·01 ppm K, in fact) or the system must be virtually dry (with
f_{H_2O} less than 10 bars).[18] This question of availability of H_2O in retrograde
reactions will be discussed later.

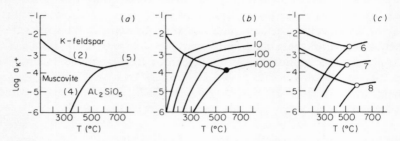

Fig 2.1 (*a*) $T-a_{K^+}$ diagram at constant pH$=7$ and $f_{H_2O}=1$ kb. Reactions
(2), (4) and (5) are discussed in the text, and intersect at a five-phase point.
Quartz and fluid are present everywhere.
(*b*) The same diagram, but showing how curves for reactions (4) and (5)
vary with changing f_{H_2O} (in bars). Note that reaction (2) is independent of
f_{H_2O}.
(*c*) The same diagram, showing how the reaction curves change drastically
with changing pH.
After Eugster, p. 116.[18]

Solid → solid reactions   Some examples of reactions involving only solid
phases are:

(*a*) *Polymorphic transformations,* such as kyanite $\rightleftharpoons$ sillimanite, in
which ΔG generally is small, especially if the reaction is *reconstructive*
(i.e. it involves breaking of chemical bonds; e.g. quartz $\rightleftharpoons$ tridymite).
Because of the small ΔG, reconstructive changes may be sluggish, and
metastable persistence of phases may occur (Chapters 3, 4). *Displacive*
(coherent) polymorphic changes involve distortion, but no breaking of
chemical bonds, so that they are rapid and unquenchable (e.g. α quartz $=$
β quartz). However, nonhydrostatic stress can affect the temperature of
these displacive reactions, as shown by experiments on the $\alpha \rightleftharpoons \beta$
transition, using oriented cores of a quartz single crystal.[12] The inversion
temperature is raised 10·6$\pm$0·4°C and 5·0$\pm$0·4°C per kb of differential
stress, during uniaxial compression normal and parallel, respectively, to the
c-axis, at a confining pressure of 3 kb. Thus, the temperature of the
$\alpha \rightleftharpoons \beta$ transition depends strongly on the orientation of the *c*-axis with
respect to the principal stress axes. Moreover, this is a thermodynamic and
not a kinetic effect, in that any phase change characterised by a reversible

transformation strain that involves a significant change of shape as well as of volume depends on *shear stress* as well as on hydrostatic stress, and shear stress can be written into the thermodynamic equations.[41]

(*b*) *Order-disorder reactions*, such as microcline $\rightleftharpoons$ orthoclase, in which certain elements (e.g. Al, in $KAlSi_3O_8$) become ordered into particular structural sites at lower temperatures and disordered at higher temperatures.

(*c*) *Mixing reactions*, such as albite and orthoclase mixing to form a homogeneous alkali feldspar solid solution at higher temperatures, and exsolving into intergrowths of two phases on cooling. Such intragranular reactions are common in compositionally complex silicates (e.g. feldspars, pyroxenes and amphiboles), and may be used to interpret subsolidus cooling histories of the minerals concerned.[45]

(*d*) *Reactions between different solids*, such as albite $\rightarrow$ jadeite $+$ quartz, a number of which will be discussed in this book. Though not as common as devaporisation reactions, they are particularly useful PT indicators in metamorphic terrains, because they are unaffected by the chemical potentials of volatile components, such as H_2O and CO_2.

Variance of Metamorphic Reactions

Recall the Phase Rule for a system in chemical equilibrium,[37] namely:

$$F = C + 2 - P$$

where $P = $ the maximum number of phases, $C = $ the smallest number of independent chemical components in terms of which the phases can be expressed, $F = $ variance of the system. The variance is the number of controlling variables that can be altered independently of the other variables without changing the phases present; this is also called the number of 'degrees of freedom' of the system.

For example, consider the system Al_2SiO_5, containing 3 phases (kyanite, andalusite, sillimanite). We could choose Al_2O_3 and SiO_2 as components, but each phase can be fully represented by one component (Al_2SiO_5), so $C = 1$. If $P = 1$:

$$F = 1 + 2 - 1$$
$$\therefore F = 2$$

So, on a PT diagram (Fig. 2.8*a*) each phase occupies an area (*divariant field*) in which P and T can be varied independently without changing the phase. However, if $P = 2$, $F = 1$, so that two phases can coexist only on a line (*univariant curve*), along which, if P is stipulated, T is automatically stipulated also (and vice versa); i.e. a two-phase assemblage in this system has a variance of one (or one degree of freedom). Similarly, if $P = 3$, only a single point (*invariant point*; 'triple' point) occurs at which the three phases can coexist; both P and T are fixed (invariable).

This is a very simple example, but the Phase Rule can be applied also to more complicated situations. Components must be chosen carefully, however. Variables are P, T, and X, where X may be the mole fraction of a volatile, or of another chemical component that does not cause the formation of an additional phase. For example, if zirconium is considered as a component in a typical silicate system, the variance would be the same as if it were ignored, because it nearly all occurs in a separate phase (zircon), which also can be left out of consideration (because the reactions we are considering do not involve zircon). In other words, an increase in C is balanced by an increase in P, so that F remains the same.

Consider a relatively simple metapelitic system consisting of SiO_2, Al_2O_3, MgO, FeO, K_2O, H_2O, and CaO ($C=7$). Consider also the following realistic biotite-forming reactions in this system: [5]

$$\text{muscovite + stilpnomelane + actinolite} \qquad (1)$$
$$\rightleftharpoons \text{biotite + chlorite + epidote} + H_2O$$

$$\text{muscovite + stilpnomelane} \rightleftharpoons \text{biotite + chlorite} + H_2O \qquad (2)$$

For reaction (1), $P=6$, so that $F=2$. If the variables are P, T, and f_{H2O} (i.e. ignoring possible variation in f_{O2}; see later), only two of these can be fixed independently. So, if P and f_{H2O} are specified, the T of the reaction is also specified; i.e. the reaction is *discontinuous*, and produces biotite of a fixed Mg/Fe ratio. On the other hand, for reaction (2), $P=5$, so that $F=3$, which means that P, f_{H2O} and Mg/Fe ratio of the reactants must be specified before T can be specified; i.e. the reaction is *continuous* ('sliding'), and will occur over a T *interval* at fixed P and f_{H2O}. As the temperature rises, (i) progressively more biotite and chlorite become stable, (ii) progressively more Mg-rich biotite becomes stable and (iii) progressively more Fe-rich muscovite and stilpnomelane become unstable. So, an isograd marked by the first appearance of biotite in an area in which both reactions (1) and (2) occur, effectively can only be based on reaction (1), which is discontinuous and involves the disappearance of the three-phase aggregate muscovite-stilpnomelane-actinolite (although two

of them can persist stably above the isograd), as well as the sudden appearance of the new phase (biotite). Reaction (2) is unsatisfactory because it is
continuous, and the first appearance of biotite in rock compositions showing this reaction will depend on the Mg/Fe ratio of the rock (and hence
the reactants). Since this ratio could be variable from one rock unit to
another in an area, the biotite 'isograd' could be similarly variable.[5] The
problems of locating a suitable biotite isograd in the Scottish Highlands
will be discussed in Chapter 4.

The above system and reactions are relatively realistic, but usually at
least minor amounts of MnO, Na_2O and Fe_2O_3 are present. Because these
components do not cause the formation of new phases, they substitute
in the phases mentioned in the above reactions, and so they may be
additional independent variables affecting the stability of the phases.

Of course, we should note also that reactions can 'slide' only if there is
a difference in, for example, the Mg/Fe ratio of reactants and products;
otherwise the reaction will be discontinuous. For example, Hollister found
that in the Kwoiek area of British Columbia the Fe/Fe+Mg ratios of
garnet and staurolite are very similar[32], so that the reaction:

$$\text{staurolite } + \text{ quartz} \rightleftharpoons \text{garnet } + \text{ sillimanite } + H_2O$$

which theoretically should be continuous, is discontinuous.

The effect of compositional variance on the PTX conditions of a particular reaction depends on the way in which the component(s) under
discussion substitute in reactants and products. The rule is that a component at a reaction site enlarges the stability field of the phase into which
it substitutes most easily. Alternatively, we can say that the addition to a
system of components consumed by a reaction will tend to decrease the
stability field of the reactants, whereas components produced by a reaction
will tend to increase the stability field of the reactants. For example, in the
sliding reaction illustrated in figure 2.2(b) Mg preferentially substitutes in
cordierite, rather than garnet, so that, with increasing Mg/Mg+Fe ratio
of the system, the stability field of cordierite is enlarged. As a result, a
hypothetical region would show the appearance of garnet at lower
pressures in Mg-poor rocks, than in Mg-rich rocks. This bulk chemical
control of cordierite-garnet relationships is emphasised by the restriction
in low-pressure metamorphism of almandine-rich garnet to rocks with
low Mg/Fe ratios; otherwise cordierite is stabilised to the exclusion of
garnet (unless Mn is present, in which case spessartine-rich garnet will
occur, owing to preferential substitution of Mn in garnet).[11]

The above discussion assumes that H_2O is present as a separate phase in the reactions. If not, P is decreased by one and F is increased by one, so that reaction (1) could become a continuous reaction under this condition, or if H_2O is a mobile component (see later).

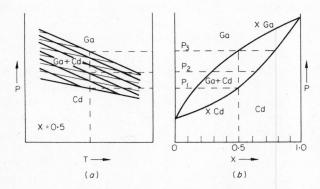

Fig. 2.2   Diagrams showing the divariant reaction cordierite $\rightleftharpoons$ garnet + sillimanite + quartz. (*a*) PT plot at constant $X = 0.5$ (where $X = Mg/Mg + Fe^{2+}$ ratio of the bulk composition of the system, as in (*b*)); showing the relatively shallow (i.e. pressure-sensitive) slopes of the boundaries of the divariant (garnet + cordierite) field and X_{Ga} and X_{Cd} contours (parallel to the upper and lower limit lines, respectively). (*b*) PX plot at constant T. At P_1 and P_3 only cordierite and garnet, respectively, are stable, whereas at P_2 both garnet and cordierite are stable, with X_{Cd} and X_{Ga} intersecting at a particular point, as shown in (*a*). After Hensen p. 199.[31]

Metamorphic Reactions and Resulting Assemblages

After completion of a reaction: $A + B \rightleftharpoons C + D$ at a particular T and P, the resultant mineral assemblage may be $C + D + A$ or $C + D + B$, depending on whether A or B is used up first. This is simply a reflection of the bulk chemical composition of the system. However, co-existence of $A + B + C + D$ implies (i) disequilibrium, (ii) P-T conditions exactly on a univariant reaction curve (rather unlikely), or (iii) P-T conditions in a divariant reaction interval (for 'sliding reactions'). The proportions of $(C + D)$ to A or B will depend merely on the bulk composition, so that mineral proportions are not used as a basis of detailed rock classification in metamorphic petrology, or as grade indicators.

Sliding Reactions at Isograds

Because many reactions in metamorphic rocks are sliding reactions, products and reactants commonly coexist over a range in grade. Isograds could still be based on the first incoming of a single index mineral, but assemblages may have greater significance and thus be better grade indicators. A good example is provided by Carmichael's mapping of isograds based on assemblages formed during a single metamorphic episode at Whetstone Lake, Canada.[9] He inferred the following reaction from observation of mineral assemblages:

chlorite + muscovite + garnet $\rightleftharpoons$ staurolite + biotite + quartz + H_2O

Because of the different partitioning of elements between reactants and products, all six phases coexist over a distance of up to 400 metres above the isograd (the 'staurolite-biotite isograd'). Note that staurolite without biotite (and vice versa) could occur first at lower grades, but the assemblage (staurolite + biotite) can occur only above the isograd. Similarly, any two of the reactants can persist stably at higher grades with or without any or all of the products, but not all three reactants. Because only (chlorite + muscovite + garnet) is eliminated by the reaction, only this three-phase assemblage is characteristic of the zone below the isograd.[9] This is an important approach, and gets us closer to basing isograds on *realistic* reactions (Chapter 4).

Effect of One Phase on the Stability of Another

The addition of another phase (B) to a particular phase (A) cannot enlarge the stability field of A, but may reduce it (by providing something with which A can react), as shown diagrammatically in figure 2.3. Similarly, the stability field of (A + B) can be reduced, but not enlarged, by the presence of phase C. Obviously, the maximum stability field of a phase will apply to rocks containing that phase on its own. Equally obviously, the stability field of (A + B) must encompass the stability fields of all assemblages of which (A + B) is a member, e.g. the assemblage (A + B + C), as shown in figure 2.7.[22, 49]

A good example has been discussed by Rosenfeld (Fig. 2.4).[44] The reactions involved are simple enough to represent on triangular diagrams (note the tie-line changes from one metamorphic zone to another),* and

*Triangular diagrams have been used extensively by metamorphic petrologists to illustrate graphically the differences between mineral assemblages from one metamorphic zone to another.[49, 53] This is helpful if the lines joining co-existing minerals ('tie-lines') change from one zone to the next in such a way as to indicate clearly what reaction has taken place. Another example is given on p. 128, using AFM diagrams, which are often useful for this purpose in metapelitic rocks.[49, 53] However, many reactions are too complicated for simple triangular diagrams (especially the ACF diagram[49, 53]), which fail to indicate what reactions have occurred. There is no point in using triangular diagrams under these conditions.

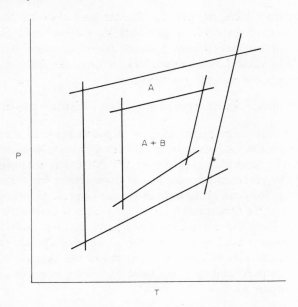

Fig. 2.3   Hypothetical stability field of mineral A (bounded by four 'break-down' or 'A-out' reactions) and the reduced stability field of assemblage (A+B).

the map shows that muscovite reacts in quartz-bearing rocks at a lower grade than it does in quartz-free rocks. This is exactly what we would expect from the foregoing paragraph, and the approximate positions of the relevant PT reaction curves confirm it (Fig. 2.5). This example emphasises the role of bulk chemical composition of the rock system. For example, we cannot talk about the 'K-feldspar isograd' in this area without specifying the rock-type, because there are two K-feldspar isograds, one in quartz-bearing, the other in quartz-free, rocks. In fact, rather than using one mineral to name a metamorphic zone, we have to use assemblages. For example, Zone A is characterised by (muscovite + quartz) and Zone C by (K-feldspar + corundum). Zone B can be determined only by observing two assemblages (one in each rock type), namely: (K-feldspar + sillimanite) and muscovite.

Criteria of Stable Coexistence of Metamorphic Minerals

In order to determine a true mineral paragenesis (Chapter 1), certain criteria must be applied. These are difficult to establish[56], but a number of microstructural features are generally used, namely: (1) common grain contacts between minerals, especially smooth, 'clean' contacts; (2) lack of evidence of one mineral replacing another (e.g. finer-grained aggregates in veinlets and along grain boundaries); (3) presence of stable shapes of grains and inclusions (Chapter 5); and (4) presence of a relatively small number of minerals, so that the assemblage obeys the Phase Rule.

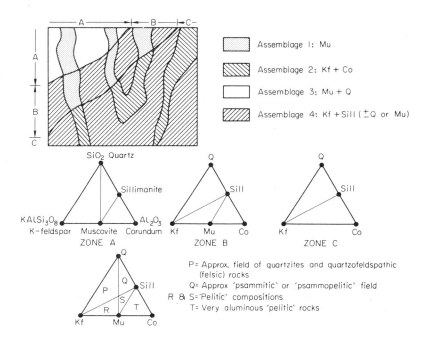

Fig. 2.4 Sketch map of a hypothetical area showing rock units and isograds for the reactions: muscovite + quartz = K-feldspar + sillimanite + vapour and muscovite = K-feldspar + corundum + vapour, together with diagrams graphically illustrating phase relationships in each zone, and a diagram showing the approximate compositional fields of relevant broad rock-types. Based on Rosenfeld, p. 4.[44] Co = corundum; Kf = K-feldspar; Mu = muscovite; Q = quartz; Sill = sillimanite; V = water vapour.

Criteria (1), (2), (3) and (4) are reasonable indicators, especially if taken together. However, criterion (2) is unreliable because minerals rimming each other can reflect an incomplete reaction (Chapter 4) and can mean also that one of the reactants became exhausted, so that the reaction stopped, leaving a partial replacement microstructure. Also, criteria (3) and (4), though important, are not sufficient indicators of equilibrium. For example, growing mineral grains are in chemical equilibrium as soon as they impinge, after which adoption of stable grain shapes occurs, so that absence of stable shapes is not necessarily indicative of chemical disequilibrium. Similarly, though systems in equilibrium necessarily obey the Phase Rule, so may systems in disequilibrium. Criterion (4) also involves the problem of how to select components in rocks.[49]

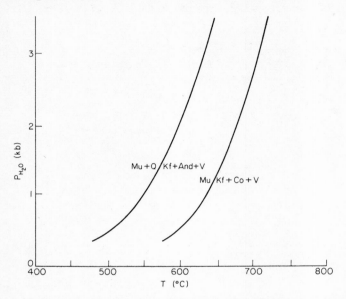

Fig. 2.5 P_{H_2O}-T curves showing the upper stability limits of (muscovite + quartz) and muscovite. After Evans (1965), *Amer. J. Science*, **263**, pp. 655, 660. And = andalusite; Co = corundum; Kf = K-feldspar; Mu = muscovite; Q = quartz; V = water vapour.

Probably the best criteria for equilibrium are:

(i) all phases of an equilibrium assemblage in contact with each other, thereby excluding phases 'shielded' from reaction, such as inclusions in another phase;

(ii) a similar composition for the *edges* of different grains of the same phase;

(iii) similar distribution of components between a pair of phases in different parts of the rock volume under discussion—this aspect of metamorphism is mentioned later in the chapter;

(iv) consistent temperatures obtained from a study of oxygen isotope ratios* between different minerals in different parts of the rock volume under discussion;

*The partition of oxygen isotopes between quartz and coexisting silicates can be used as a geothermometer, quartz having a high, constant O^{18}/O^{16} ratio at all temperatures.[15] The isotope ratios vary significantly with

(v) stable shapes of grains and inclusions of the minerals concerned, as discussed above.

Domains of Local (Mosaic) Equilibrium[4,48]

The attainment of chemical equilibrium through large volumes of rock is restricted by: (*a*) original chemical heterogeneities, (*b*) limitation of diffusion to and from reaction sites, (*c*) slow reaction rates, which could be related to (*b*), and (*d*) zoning of some of the minerals formed. All these features reflect relatively slow diffusion rates under the prevailing conditions. As pointed out by Blackburn, the problem is to assign a volume of equilibration to a particular phase with respect to a given element, under specified metamorphic conditions.[4] His electron microprobe analyses showed that chemical analyses of minerals mechanically separated from a hand specimen really represent average compositions. He found that domains of spatial equilibration of Mg and Fe in garnet range from only a few millimetres to a few centimetres across, and that the domains tend to be elongate parallel to the foliation and lineation of the rock. Spatial equilibration of Fe tends to be more extensive than for Mg, the Mg equilibration domains tending to be much more regular in their orientation, which possibly reflects the higher diffusional mobility of Fe relative to Mg. The domains appear to grow larger with increasing metamorphic grade; for example, in a pyroxene granulite, Blackburn found equilibration, parallel to the foliation, over areas of almost hand specimen size.[4]

A detailed chemical and microstructural study of some inferred reactions involving the breakdown of staurolite (Chapter 4) has led Kwak to suggest that in many prograde metapelitic rocks the reactions involve local systems on the scale of only a few grains (e.g. one staurolite porphyroblast and the reaction products that partly replace it as a 'corona').[39] On this scale the systems appear to be open, because (i) the same minerals involved in different microstructural situations in the same 'rock' (say, on the scale of a single thin section of standard size) have different chemical compositions, and (ii) a comparison of the masses of reactant staurolite and inferred products in the replacement corona consistently reveals a chemical imbalance, indicating gains and losses on the scale of the local system. One of the relevant reactions is discussed in more detail in Chapter 4.

So it appears that diffusion in metamorphism, even in rocks of relatively high grade, may be limited (Chapter 3), so that we should investigate carefully the possibility of local equilibrium in all metamorphic rocks.

---

temperature, especially between 100 and 500°C, but are virtually insensitive to pressure.

Calculation of the Topology of Metamorphic Phase Diagrams (Schreine-makers' Analysis)

A relatively old approach of Schreinemakers in the 1920s has been taken up recently.[57] Using the Phase Rule, the topology of any phase diagram can be constructed if the number of phases and the number of components are stipulated. Nowadays, computers will do the job for quite complicated systems. Once the topology is known, the positioning of invariant points (in P-T space, for example) and the slopes of univariant lines can be obtained from experiment, or slopes can be calculated thermodynamically.

The method is useful (1) as a check on experimentally determined and petrologically inferred phase diagrams (which must conform with the Phase Rule topology), and (2) as a general indication of mineral relationships in the absence of experimental information. An example of a simple Schreinmakers' analysis for a system with one invariant point is shown in figure 2.6.

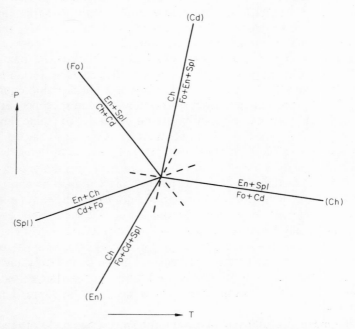

Fig. 2.6 Arrangement of all possible univariant curves around the invariant point involving the phases: forsterite (Fo), enstatite (En), spinel (Spl), cordierite (Cd) and chlorite (Ch) in the system: $MgO-Al_2O_3-SiO_2$. The curves are labelled by the absent phase, which is also absent from the fields on either side of the reaction line. After Fawcett and Yoder (1966). *Amer. Mineralogist*, **51**, p. 365.

Because the Phase Rule permits the number of phases coexisting stably at an invariant point to exceed the number of components by no more than 2, the representation of the stable coexistence of more than $(C+2)$ phases on a P-T grid must contain more than one invariant point. This collection of invariant points, joined by a 'net' of univariant lines, is called a 'multi-system'.[36] Multisystems also are subject to certain topological restrictions, so that calculations enable graphical representations of P-T grids that are permissible in terms of these restrictions.[13, 58, 59, 61] Thermodynamic data may enable selection of the most likely configurations. The resulting grids (e.g. Figs 4.5, 4.12) can be valuable indicators of reaction trends with changing P-T conditions, especially in the absence of sufficient experimental information.

Calculation of Slopes of Reaction Curves

As we have seen previously, the fundamental thermodynamic equation of a reaction at equilibrium is:

$$d(\Delta G) = O = (\Delta V)dP - (\Delta S)dT$$

$$\therefore \left(\frac{dP}{dT}\right)_{\Delta G = O} = \frac{\Delta S}{\Delta V}$$

Therefore, if we know the changes in entropy and molar volume involved in a reaction, we can calculate the slope of the equilibrium curve on a P-T diagram. To do this we need accurate values of entropies and molar volumes of reactants and products, as functions of P and T. Entropies of minerals are best measured directly by using calorimeters, which measure heat capacity at different temperatures. This work is beginning to be done for common minerals, but, if direct measurements are not available, less reliable estimates can be made in various ways.[43, 49] Molar volumes can be obtained from refined X-ray crystallographic data. Tabulated thermodynamic data are available for many minerals, but are subject to continued improvement.[43]

For reactions involving only solid phases, changes in ΔS and ΔV with changing P and T are small enough to neglect for approximate calculations, so that $\Delta S / \Delta V$ is constant, and solid-solid reactions generally are shown as straight lines on P-T grids. Provided a good experimental reversal $(\Delta G = O)$ has been made at one P-T position, the rest of the reaction line can be drawn by extrapolation. However, many minerals have ordered distributions of certain elements in certain structural sites at lower temperatures, and disordered distributions (which increase configurational entropy) at higher temperatures. In silicates, a common example is the tendency for Si and Al to occupy preferred tetrahedrally co-ordinated sites in the ordered state. Strictly, therefore, S due to disorder should be taken into account when calculating slopes of reactions.

In reactions involving a fluid phase, ΔS and ΔV are affected by changing P and T, so that $\Delta S/\Delta V$ is not even approximately constant. The effect of the fluid can be separated from that of the solid phases in the following way (taking a dehydration reaction as an example):

$$d(\Delta G) = O = (\Delta V_s)dP - (\Delta S_s)dT + \Delta G_{H_2O}$$

Usually ΔV_s (i.e. ΔV of the solid phases) is assumed to be independent of P and T, and ΔS_s is assumed to be independent of P, only small errors being introduced. The variation of ΔS_s with T and of G_{H_2O} with P and T must be taken into account, and, once again, thermodynamic compilations help in obtaining the necessary data for most purposes.[7, 43] Therefore, an approximate equation for calculating the slope of a dehydration reaction at P_1T_1 is

$$\left(\frac{dP}{dT}\right)_{\Delta G=O} = \frac{\Delta S}{\Delta V} = \left[\frac{(\Delta S_s)_{T_1} + [(S_{H_2O})_{P_1T_1}]n_{H_2O}}{(\Delta V_s)_{1\,bar,\,298°K} + [(V_{H_2O})_{P_1T_1}]n_{H_2O}}\right]$$

where n_{H_2O} represents the number of moles of water produced in the reaction.[47, 49]

More accurate calculations use thermal expansion and compression data to estimate ΔS_s as a function of P ($\approx \partial V/\partial T$ at 1 bar) and ΔV_s as a function of P ($\approx \partial V/\partial P$ at 298°K).[33, 51] They also take into account the effect of Al/Si disorder on ΔS_s as a function of T.[51] Then, if we have a reliable, experimentally determined reversal ($\Delta G=O$) at P_1T_1, integration (by computer) of the following equation gives P-T co-ordinates at other points (e.g. P_2T_2) along the equilibrium curve:[33, 51]

$$\Delta G = O = \int_{P_1}^{P_2} (\Delta V_s)dP - \int_{T_1}^{T_2} (\Delta S_s)dT + (G_{H_2O})_{P_2T_2} - (G_{H_2O})_{P_1T_1}$$

Free energy data derived from hydrothermal experiments can also be of great assistance in thermodynamic calculations.[1, 3, 20, 24, 60]

Examples of calculated reaction lines that agree well with experimental work are shown in figure 2.7. Careful calculations can (i) reduce the need for much expensive, time-consuming experiment, and (ii) act as a check on experiment.

Slopes of metamorphic reactions on P-T diagrams may vary considerably. Most devaporisation reactions have steep slopes (Figs 2.5, 2.8*b*), and so the reactions are mainly sensitive to variation in temperature. This

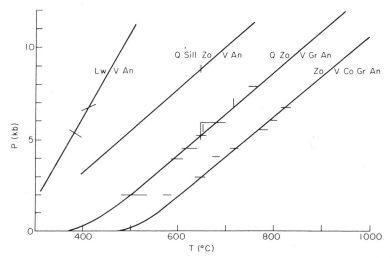

Fig 2.7   Some calculated reaction curves in the system $CaO - Al_2O_3 - SiO_2 - H_2O$ by Wall and Essene,[51] p. 700, showing the excellent agreement between calculations and experimental reversals. Compare the stability limits of Zo, (Zo+Q) and (Zo+Q+Sill). An=anorthite; Co=corundum; Gr= grossular; Lw=lawsonite; Q=quartz; Sill=sillimanite; V=water vapour; Zo=zoisite.

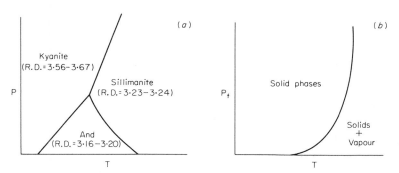

Fig. 2.8 (*a*) Schematic phase relationships in the system Al_2SiO_5 showing relative densities; (*b*) typical shape of a devaporisation reaction curve on a P_f-T diagram.

means that the effect of ΔS exceeds that of ΔV in these reactions. Solid-solid reactions may have steep (Fig. 2.8*a*) or shallow (Fig. 2.2) slopes on P-T diagrams. Shallow slopes indicate a predominance of ΔV over ΔS.

Such reactions are valuable in metamorphic areas, because they intersect the steeply sloping, more common devaporisation reaction curves, and thereby provide more divariant fields on a P-T 'grid' (Fig. 2.14). The Al_2SiO_5 system has curves with both positive and negative dP/dT slopes (Fig. 2.8*a*), which can be related to the ΔV involved in the polymorphic changes.

If two reactions have markedly different slopes (e.g. a devaporisation and a solid-solid reaction), the metamorphic zoning based on these reactions can be reversed from one area to another if the geothermal gradients are sufficiently different (Fig. 2.9).

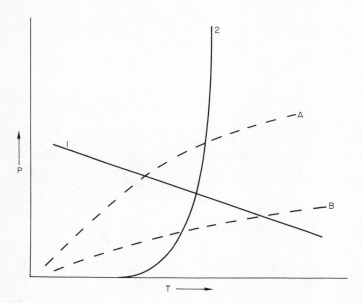

Fig. 2.9   P-T plots of two hypothetical reactions (1 and 2) in rocks of different composition, on which are superimposed the geothermal gradients of two hypothetical areas (A and B), showing zonal sequence $1 \rightarrow 2$ in area A, and $2 \rightarrow 1$ in area B.

Devaporisation curves change their slopes, especially at lower pressures, because the fluid phase becomes less compressible at higher pressures, so that ΔV decreases with increasing P (Figs 2.8, 4.5, 4.12, 4.17, 4.19). An extreme change of slope is shown by the reaction: analcite + quartz $\rightleftharpoons$ albite + H_2O (Fig. 2.10). Because of the open structure of analcite, ΔV of the solid phases in this reaction is negative, which means that ΔV of the whole reaction is positive only up to a P_{H_2O} of around 1 kb, above which the curve has a negative slope. At pressures of more than 30 kb, most dehydration reactions have negative slopes.[21]

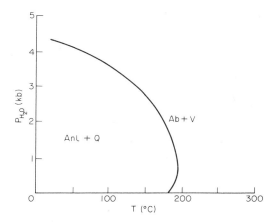

Fig. 2.10 P_{H_2O}-T curve for the reaction analcite + quartz $\rightleftharpoons$ albite + H_2O determined partly experimentally, partly theoretically by Campbell and Fyfe (1965). *Amer. J. Science*, **263**, p. 813.

Experimental Location of Reaction Curves

The following main types of apparatus are used: *

Piston-cylinder apparatus The sample is enclosed in a small graphite furnace (the temperature being controlled by a thermocouple), which itself is enclosed in a weak solid. Commonly the solid is talc if water is desired (since the talc dehydrates on heating), or boron nitride if water is to be excluded. This whole assembly is confined in a large steel cylinder and pressure is exerted by a rammed piston. The pressure is transmitted to the sample by the weak solid. This kind of apparatus can go to very high temperatures and pressures (commonly 1 600°C and 50 kb). A negligible temperature gradient is present in the sample (±5°C) but, more important, the pressure is known only approximately (±10%). This is because of (i) friction between the cylinder and piston (which can be corrected only approximately), and (ii) the strength of the solid pressure-medium. These factors also mean that the pressure on the sample is not completely hydrostatic. Such shortcomings do not matter so much if the ΔG of the reaction is large (as in most devolatilisation reactions), but they

*Detailed information on experimental procedures can be obtained from a recent book by Edgar, A. D. (1973) *Experimental Petrology. Basic Principles and Techniques*. Oxford: Clarendon Press.

can make the results too inaccurate to be useful if the ΔG is small (as in many polymorphic reactions). Quenching can be achieved by turning off the current, and normally is so rapid that any further reaction in metamorphic systems is unlikely to be caused.

Hydrothermal apparatus[16]   The sample is sealed in metal foil (generally platinum) and placed in a strong metal pressure vessel ('bomb') into which fluid is pumped, producing a truly hydrostatic pressure on the sample. Temperature is provided by a wound resistance furnace surrounding the bomb, and is measured by a thermocouple. Rapid quenching is achieved by compressed air. Variation in f_{H_2O} in the sample contained can be obtained by the addition of known amounts of silver oxalate (which decomposes, thereby diluting the water with carbon dioxide). Where the valency of transition elements (especially iron) is to be controlled, the oxygen fugacity can be held at a required level by surrounding the charge capsule by an oxygen buffer assemblage plus water, which itself is enclosed in a capsule effectively impermeable to hydrogen.[17] The oxygen buffer assemblage and water equilibrate at a certain f_{O_2} and f_{H_2}, and, because hydrogen can diffuse through the platinum into the inner capsule, the water there dissociates until the same f_{O_2} as that in the outer capsule is achieved.

Reversal of Experimental Reactions

In order to locate a reaction boundary accurately, reactions must be run in the reverse as well as the forward direction, allowing the possible equilibrium positions to be bracketed within the limits of experimental detection (Fig. 2.11). If the reaction is run in the forward direction only, kinetic difficulties may cause the reaction to overstep the reaction point (e.g. the right-hand points of Fig. 2.11). In fact, some overstepping always occurs in processes (including chemical reactions) involving nucleation of a new phase or phases (Chapter 3).

Discrepancies Between Calculated and Experimentally Determined Reaction Curves

Turner has discussed the necessity of checking experimental work against thermodynamic calculation, and vice versa.[49] Discrepancies are relatively common owing to kinetic problems in the experiments, such as use of high-energy starting materials as discussed in Chapter 3, but good agreement has been achieved in some systems (Fig. 2.7). Wherever possible, I will refer to systems and reactions with a fair measure of agreement between the two approaches, but both are subject to constant improvement.

Open Systems in Metamorphism[25,34,35,36,47,48]

Petrological evidence indicates that devaporisation reactions are common in prograde metamorphism. Removal of the fluid produced (regardless of the mechanism, which is discussed in the next section) implies that

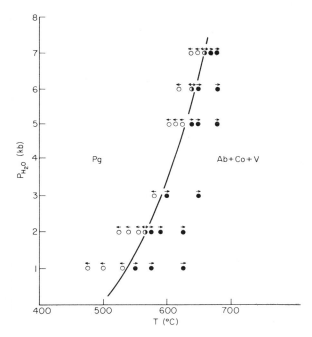

Fig. 2.11 Univariant equilibrium curve for the reaction: paragonite $\rightleftharpoons$ albite +corundum+ H_2O on a P_{H_2O}-T diagram, as determined experimentally by Chatterjee (1970). *Contribs. Mineralogy & Petrology*, **27**, p. 251. Solid circles represent breakdown of paragonite (growth of albite+corundum), whereas open circles represent breakdown of albite+corundum (growth of paragonite). Half-filled circles represent no apparent reaction. Arrows show the directions of reaction.

the rock system under discussion is open to some components. The same applies to the addition of fluid in the hydration of formerly anhydrous (e.g. some igneous) rocks. Therefore, some components of a rock system may be *perfectly mobile* (i.e. their chemical potentials are defined externally and not by their concentration in the system), whereas others may be *immobile* (i.e. their chemical potentials are defined by their concentration in the system itself). If an aqueous fluid moves through a rock, various dissolved components (e.g. MgO, CaO, K_2O, Na_2O, CO_2) may become mobile. However, if the movement of a component (either in aqueous solution or in diffusing 'independent' form) is too slow relative to the rate of a particular reaction taking place in the rock, the component may cease to become mobile, by having its chemical potential fixed by the reaction concerned.[28,50]

If a system is open to a component, i, there is a tendency towards the establishment of equilibrium, such that the chemical potential of i (μ_i) is equalised in both the system and its immediate surroundings (external reservoir of component i). So, μ_i inside the rock system is controlled by μ_i in the reservoir; i.e. it is imposed on the system from outside. A distinction can be made between 'volatile' mobile components (i.e. those that would occur as a fluid if pure, under the prevailing P-T conditions, such as H_2O and CO_2), and 'non-volatile' mobile components (e.g. KCl), but both can be treated the same way in terms of the theory.[50]

As stated previously, for a completely closed system at equilibrium, $d(\Delta G)=O=(\Delta V)dP-(\Delta S)dT$, which represents a condition of *univariant* equilibrium in P-T space (i.e. if P is fixed then so is T, and vice versa, in order for the system to remain at equilibrium somewhere along the univariant line).

However, for a system open to a component, i, at equilibrium, $d(\Delta G)=O=(\Delta V)dP-(\Delta S)dT+(\Delta n_i)d\mu_i$, where ΔV and ΔS refer to changes in the rock system and n_i=number of moles of i in the external reservoir. This represents a condition of *divariant* equilibrium (in PTμ space, as discussed later*), since two of the three variables (P, T and μ) can be chosen independently and still remain on the equilibrium surface.

If two or more components (i, j, etc.) are mobile, the *trivariant* (i, j) or *multivariant* (i, j, ... m) equilibrium is expressed by:

$$d(\Delta G)=O=(\Delta V)dP-(\Delta S)dT+\Sigma_{i,\,j,\,\ldots\,m}\left\{(\Delta n_{i,\,j,\,\ldots\,m})d\mu_{i,\,j,\,\ldots\,m}\right\}$$

These equations can be represented mathematically, not graphically, as reactions of $(m+1)$–variant equilibrium in $(m+2)$–dimensional space,

---

*Later discussions actually refer to PTX space, rather than PTμ space, because we are trying to express relationships in systems involving more than one mobile component, so that X_i represents the mole fraction of i in a group of different mobile components. Use of X is preferable to μ under these conditions because it can be measured.[26] As mentioned previously, the relationship between X_i and μ_i is: $\mu_i=\mu_i{}^\circ+RTlnX_i$ (if the mixture of mobile components is ideal) or $\mu_i=\mu_i{}^\circ+RTlnX_i\gamma_i$ (if the mixture is not ideal, where γ_i is the activity coefficient of i in the mixture). Note that in a system involving only one mobile component (e.g. pure H_2O), μ_i and n_{H_2O} (concentration of i, expressed as number of moles) are not necessarily equivalent.[47]

where m is the number of mobile components. So we see that the presence of each mobile component adds another variable to the system, so that, for an open system, the Phase Rule becomes:

$$F = C_{\text{immobile}} - C_{\text{mobile}} + 2 - P$$

This is known as Korzhinsky's Phase Rule, or the Korzhinsky–Thompson Phase Rule, and implies that, as the number of mobile components in a rock increases, the number of phases decreases.[35] For example, in certain banded calcsilicate rocks inferred to have been formed by metasomatic processes, the number of phases in an assemblage decreases by one for each component inferred to be mobile in the band concerned.[50] It follows that the end-product of metasomatic processes involving increasing solubility (and hence mobility) of rock components in a compositionally zoned rock sequence formed by progressive metasomatism should be a rock with relatively few minerals.[35] This has been used as an argument against the formation of granitoid rocks (which generally have at least four major minerals) as the end-products of progressive metasomatism.

Fluid Phase in Metamorphism

Metamorphic petrologists commonly assume that a fluid phase is present always during metamorphism. In order to obey the Gibbsian definition of a 'phase' this must be, potentially at least, mechanically separable from the solid phases. Intuitively, minor amounts of such a fluid phase would be expected in porous (especially sedimentary) rocks undergoing initial metamorphism, except for very deeply buried rocks (*ca.* 30 000 feet?). However, its presence must be questioned with regard to the metamorphism of igneous rocks (which begin their metamorphism as space-filling crystalline aggregates), and with regard to the medium and high-grade metamorphism of any rock, in which grain shapes typically indicate adjustment under the influence of solid interfacial energies (Chapter 5). A quotation from Ralph Kretz is apposite, namely: 'a control on grain shape by interfacial energy . . . is incompatible with an intergranular fluid, the presence of which is commonly assumed.'[38] Of course, strictly this applies only to the final stages of grain boundary adjustment, and does not exclude the presence of a fluid phase in the pre-impingement stage of mineral growth (i.e. while the stable assemblage was being formed). Moreover, aggregates undergoing deformation may contain fluid-filled voids produced by grain boundary sliding (Chapter 6).

Arguments in favour of a fluid phase during metamorphism are:

(i) The common occurrence of minute fluid inclusions (bubbles) in

metamorphic minerals. These could have been trapped during crystal growth in the presence of a fluid phase, or they could have been produced by heating of formerly structurally bound water (in hydrolysed bonds), as has been achieved experimentally in quartz. This release of hydrogen-bonded hydroxyl ions to form a true fluid phase of water bubbles could have occurred well after impingement of stable metamorphic grains (Chapter 5), and so may not reflect the presence of a former fluid phase. Even if primary in origin, fluid inclusions may indicate the nature of earlier, rather than later, stages of metamorphic conditions. Electron micrographs of some naturally and experimentally deformed aggregates commonly show bubbles occurring locally along grain boundaries; some of these may represent voids opened up during deformation.

(ii) The occurrence of porphyroblasts with well-formed, low-energy crystal faces, similar to those that would be formed by growth in a fluid phase (Chapter 5). However, perhaps this may be achieved by adsorption of a relatively large number of foreign ions (probably including hydroxyl) in the solid/solid interface, thus markedly lowering its energy and thereby permitting rational faces to form.[38] Perhaps strong adsorption on mica$_{(001)}$/tectosilicate interfaces is possible also, and may be an explanation of the remarkable persistence of mica (001) planes in metamorphic rocks (Chapter 5).

(iii) The growth of many porphyroblasts and the development of metamorphic 'segregation' layering implies transport of material over distances much larger than the scale of a single grain. Diffusion in a fluid phase is much more rapid than anhydrous intergranular solid diffusion (Chapter 3).

Suggestions have been made that fluid (especially 'water') may be present as 'surface films' (intergranular films), which could increase reaction rates, so that most metamorphic reactions could take place by means of these films.[22] They are postulated to be no more than 'a few molecules' in thickness. However, nobody knows how continuous they are, or their exact nature. Presumably, a very large concentration of adsorbed water would be necessary for the films to be called correctly a hydrous 'phase'. Perhaps they should be regarded as essentially solid,[47] namely as water-rich grain boundaries. They should contain a large proportion of hydrogen bonding, which would assist diffusion greatly, owing to the considerably reduced necessity for breaking strong oxygen-cation bonds. So, diffusion in water-rich grain boundaries, rather than a separate intergranular water phase, may be adequate for metamorphic reactions, given the long times involved, at least on the scale of a few grains.

Greenwood has shown experimentally that water can diffuse against a confining pressure gradient through a non-porous aggregate of brucite to react with periclase and produce more brucite.[25] P_{H_2O} was much less than P_{load}, and was too low to force open pores in the brucite. Therefore, it must have moved by diffusion, presumably mainly along brucite grain boundaries. This shows the general feasibility of water diffusion in metamorphic reactions, but quantitative data are lacking.

Local minute pockets or pores rich in true water should not be excluded as a possibility, but the presence of a continuous film of a water phase (especially one capable of independent flow relative to solid grains) is highly suspect in extensively recrystallised rocks. In this connection, we should remember deformation experiments that show the remarkable degree of weakening (commonly through local zones of embrittlement) of rocks in which water is suddenly released (e.g. the dehydration of serpentinite under axial compression).[42] Such embrittled zones may occur locally in metamorphic rocks, but mostly these are not evident, unless one wants to speculate that some foliations are produced by the parallel alignment of these zones during escape of water produced in dehydration reactions (followed by recrystallisation to remove any evidence of cataclasis).

A fairly realistic picture may be that prograde metamorphism of hydrous aggregates involves the gradual expulsion of hydroxyl ions which (i) adsorb and diffuse through grain boundaries and/or (ii) collect and move as mobile bubbles during deformation (although presumably this would not apply to post-tectonic or most contact metamorphism). The increasingly hydrated interfaces would permit increased diffusion of other ions as well, thus facilitating reactions. But the essential character of a grain boundary (Chapter 5) apparently is not lost; if it were, we would not expect to see so much evidence of microstructural adjustment to minimise solid-solid boundary energies. Presumably this *grain-boundary hydration* is also a mechanism for transferring water from fractures to reaction sites in anhydrous igneous rocks during low-grade metamorphism, and relatively anhydrous high-grade metamorphic rocks during retrograde metamorphism.

Probably the concentration of hydroxyl in grain boundaries would gradually decrease with increasing metamorphic grade; and high-grade rocks tend to show the greatest adjustment of grain boundaries to a low-energy configuration. However, in the early stages, especially before the complete impingement of stable grains, the minimisation of chemical free energy overshadows the minimisation of interfacial free energy. The grain boundaries should be particularly hydrous, and this may permit the formation of porphyroblasts with rational crystallographic boundaries. Once formed, there would be little tendency for them to be removed during later grain adjustment, owing to their already low interfacial energy (Chapter 5).

The main contribution of hydrous grain boundaries would be to increase reaction rates (as well as to assist deformation, especially by grain boundary 'sliding'). A possible control of the rates of dehydration reactions could be the rate at which hydroxyl could diffuse through grain boundaries away from reaction sites. If this duffusion rate were slow enough, the reaction could be constrained to occur on the reaction curve determined experimentally in the presence of a true water bulk phase. This explanation is an alternative to the more common one involving a build-up of pressure of a fluid phase in the rock, as discussed later.

The trouble with the foregoing explanation is that relatively large quantities of water must be removed from a pelitic sediment during prograde metamorphism, and we don't know enough about the diffusion rates of hydroxyl in silicates to say whether or not diffusion would be effective in removing all this water in the time available. On the other hand, outward flow of a true water phase at medium and high metamorphic grades also meets difficulties in terms of grain-boundary configuration, unless the mechanism involving embrittlement by water-weakening is effective, or unless motion of fluid-filled, transient voids in a deforming aggregate is adequate to remove all the fluid. I think that Atherton's suggestion should be considered, namely that the expulsion of free water from pelitic sediments takes place early in the metamorphic history, when deformation is most active, porosities are high and perhaps 'dewatering'- from place to place, depending on the degree of compaction and deformation, and the loss of water eventually would be stopped by reduced porosities, lessening of deformation and growth of metamorphic minerals. In this way, distribution of *free* water for the region would be fixed at an early stage, and the metamorphic assemblages that result from the continued increase in temperature would be determined partly by this water distribution. However, dehydration reactions releasing *combined* water occur also at higher metamorphic grades, so that at least some of the water removed from a metapelitic rock must take place after the initial distribution of free water has been established.

Pressure of Fluid Phase

At high crustal levels, where rocks are strong (at low temperatures and under small confining pressures, which conditions favour increased strength, as shown by many experiments), a fluid phase may exist in porous rocks, with $P_{fluid} < P_{load}$ ($P_{confining}$, P_{total} P_{solids}). As compaction and the degree of metamorphism increase, pores would tend to become fewer, induced cleavages are being formed.[2] The amount of water lost would vary smaller and disconnected, so that P_{fluid} would tend to increase, and eventually the situation may become: $P_{fluid} = P_{load}$. Many metamorphic petrologists believe that this is the most common situation in metamorphism,[49] except in largely dehydrated rocks of the granulite facies, in which P_{fluid} may be less than P_{load}, if a true fluid phase is indeed present (see above).

If $P_{fluid} > P_{load}$, fluid would be forced out of the rock, which has been suggested as a mechanism for removing volatile products in devaporisation reactions.[49, 53] Raising the pressure of the pore fluid tends to force grains apart and so counteracts the confining pressure. It has the effect of reducing the 'effective pressure' on the mineral grains. However, once P_{fluid} exceeds P_{load}, the rock can lose cohesion easily[42] (e.g. by the local operation even of small differential stresses), so it is very unlikely that rocks could sustain this situation for long periods. Therefore, most metamorphism, if a fluid phase is present, probably takes place under conditions of $P_{fluid} \leqslant P_{load}$.

Effect of Fluid Pressure on Reaction Curves

In devaporisation reactions, the fluid produced could be removed from the reaction site to varying degrees, depending on diffusion rates. If the system is closed to the fluid, P_{fluid} will approach or equal P_{load}, and the appropriate equilibrium reaction curve will be similar to curve (i) of figure 2.12.* This is the situation in most laboratory experiments on devaporisation reactions, and may be common also in some natural situations, in which slow diffusion of volatile products may constrain the reaction to proceed on the univariant curve for $P_{fluid} = P_{total}$. On the other hand, if the fluid escapes readily from the reaction site, P_{fluid} (or f_{fluid}) becomes small, so that the reaction will proceed at lower P-T conditions, so as to attempt to produce still more fluid. The extreme situation of all fluid escaping instantly from the reaction site is represented by reaction (ii) of figure 2.12, which has a negative slope, because the volume of fluid has been neglected in the calculation. All situations between the two extremes are theoretically possible (Fig. 2.12). The same arguments apply to both dehydration (Figs 2.14, 2.15, 2.16) and decarbonation (Fig. 2.12) reactions. The effect is shown clearly on T-X_{CO_2} diagrams (Figs 2.13, 2.16).

A probable natural example of this effect occurs in drill holes in the Taringatura geothermal area, New Zealand, where relic zones of

---

*This diagram actually refers to a H_2O-CO_2 mixture, but the principle is the same, as discussed in the next section.

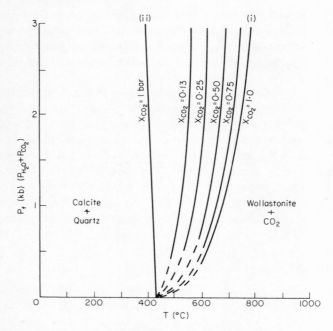

Fig. 2.12   Curves of univariant equilibrium for the reaction: calcite + quartz ⇌ wollastonite + CO_2 as functions of P_f ($= P_{H_2O} + P_{CO_2}$) for various compositions of the fluid phase (expressed as X_{CO_2}). After Winkler, p. 35.[53]

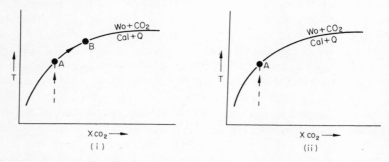

Fig. 2.13   Diagrammatic T-X_{CO_2} plot of the reaction: calcite + quartz ⇌ wollastonite + CO_2, showing two extreme situations, namely: (i) *CO_2 immobile*, so that, as calcite + quartz begin to react at A, CO_2 is released, X_{CO_2} increases and the reaction occurs with increasing T and X_{CO_2} until one of the reactants is exhausted at B; and (ii) *CO_2 perfectly mobile*, so that the reaction proceeds at a fixed T (lower than for (i)), namely at A.

heulandite occur in the higher temperature laumontite zone (zeolite facies, Fig. 1.5). The heulandite occurs in rocks of low permeability (from which water could not escape readily, so that the heulandite was inhibited from dehydrating to laumontite), whereas the laumontite occurs in more porous rocks (from which water could escape relatively easily).[22] This is another indication of the dependence of metamorphic 'grade' on the activity of volatile components (cf. Chapter 1).

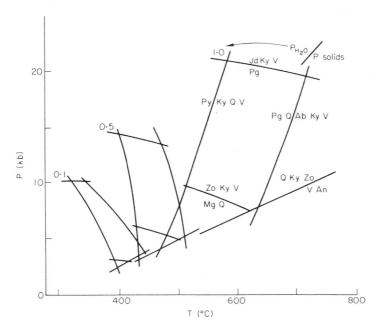

Fig. 2.14 Calculated stabilities of some assemblages involving paragonite (Pg), zoisite (Zo), kyanite (Ky), quartz (Q) and water (V) for $P_{H2O}=P_{solids}$, $P_{H2O}=0.5\ P_s$ and $P_{H2O}=0.1\ P_s$. Calculations by Wall and Essene, p. 700.[51] Ab=albite; An=anorthite; Jd=jadeite; Mg=margarite; Py=pyrophyllite.

Effects of Composition of Fluid Phase (Fugacities of Mixed Volatile Components) on Reaction Curves

Even in an O-H fluid, dissociation causes the presence of molecular species such as H_2 and O_2, as well as H_2O. Therefore, equilibria involving an aqueous fluid and silicates containing elements of variable valency (especially Fe) are affected by variation in f_{O2}, as well as f_{H2O}, P and T. This variation could be caused by local reactions in closed systems, or by concentrations in external reservoirs for open systems. The effect of oxidation in metamorphic reactions will be considered in the next section.

Furthermore, though 'water' generally is the most common fluid phase (mobile component) in pelitic and mafic metamorphic rocks, other com-

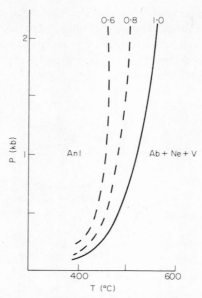

Fig. 2.15   Differing positions of the curve for the reaction: analcite $\rightleftharpoons$ albite +nepheline+H_2O, for differing mole fractions of H_2O in the fluid phase, the dilutant being argon. After Greenwood, p. 3942, © by American Geophysical Union.[25]

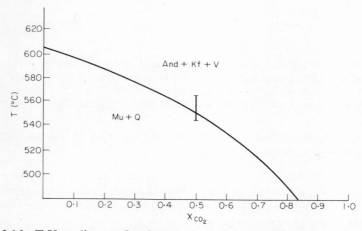

Fig. 2.16   T-X_{CO_2} diagram for the reaction: muscovite+quartz $\rightleftharpoons$ K-feldspar +andalusite+H_2O at $P_{fluid}=2$ kb, showing a calculated curve and an experimentally determined equilibrium bracket. The curve was calculated from Greenwood's equation for dT/dX_i (see text), where $i=H_2O$, using values for ΔS of the reaction based on averages of the available thermodynamic data. After Kerrick, p. 952.[33]

ponents may also occur, especially CO_2 in and near carbonate-rich rocks. The CO_2/H_2O ratio (mole fraction) of a fluid phase can vary in space and time within an area, and between different areas. Therefore, the simple situation involving only one volatile component, discussed above, is unlikely to be applicable to most metamorphic situations. In terms of P, $P_{fluid} = P_{H_2O} + P_{CO_2}$, so that the pressure of each component is less than the total fluid pressure (e.g. $P_{H_2O} < P_{total}$). More correctly, the fugacity of H_2O is diminished by the presence of other fluid components, so that the $P_{H_2O} - T$ reaction curves for dehydration reactions are moved to lower temperatures (even though CO_2 takes no part in the actual reaction). The effect has been clearly demonstrated by Greenwood who added various proportions of an inert fluid (argon) to water, thereby changing the position of the analcite breakdown curve (Fig. 2.15).[25] The same reasoning applies if we are referring to mobile (volatile) components of water and carbon dioxide in the absence of a true fluid phase.

The most spectacular natural effects of fluid phase composition can be seen in reactions involving carbonates.[26, 27, 55] The effect of dilution of H_2O by CO_2, and vice versa, can be shown on a PTX diagram (Fig. 2.17), which is Wyllie's 'petrogenetic model'.[55] The P-T curves for $P_{H_2O}=P_{total}$ and $P_{CO_2}=P_{total}$ of two simple devaporisation reactions are shown on opposite faces of the three-dimensional block, but each curve slopes down to lower temperatures with progressive dilution of the fluid phase by the other component.

In two dimensions, the chemical potential of one component (e.g. μ_{CO_2}) can be plotted against the other (e.g. μ_{H_2O}) in a 'Korzhinsky diagram'[36] (e.g. Fig. 4.13). However, a TX diagram (where X is the mole fraction of the mobile component being considered, other components being regarded as immobile) often is more useful, because it uses measurable variables.[26] The TX slope of an equilibrium boundary for a reaction at constant P involving a binary fluid phase (assuming ideal mixing) is:[26]

$$\left(\frac{\partial T}{\partial X_j}\right) = \frac{RT}{\Delta S}\left(\frac{v_j}{X_j} - \frac{v_i}{X_i}\right)$$

where X_i = mole fraction of component i in the fluid and v_i is the coefficient of i in a reaction such as: solid reactants $\rightleftharpoons$ solid products $+v_i+v_j$ (provided the equation is written so that $v_i+v_j=1$).

This can be applied to reactions involving H_2O and CO_2, if we make $H_2O=X_i$ and $CO_2=X_j$. For a reaction such as tremolite $\rightleftharpoons$ 2diopside + 3enstatite+quartz+H_2O, $v_{CO_2}=O$ and $\partial T/\partial X_{CO_2}$ is negative for all values of X_{CO_2} (Fig. 2.18). Similarly for the reaction: magnesite $\rightleftharpoons$ periclase+CO_2, $v_{H_2O}=O$ and $\partial T/\partial X_{CO_2}$ is always positive (Fig. 2.18).

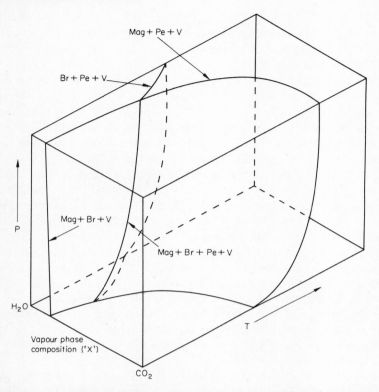

Fig. 2.17   Three-dimensional P-T-X diagram of the system $MgO\text{-}CO_2\text{-}H_2O$, showing three divariant (three-phase) reaction surfaces meeting in a univariant (four-phase) reaction line, and dividing the diagram into three trivariant (two-phase) volumes, namely $MgO+V$ (top-right), $MgCO_3+V$ (front) and $Mg(OH_2)+V$ (rear). The P-T curve on the front face is the reaction: $MgCO_3 \rightleftharpoons MgO+CO_2$ at $P_{CO_2}=P_{total}$, and the P-T curve on the back face is the reaction: $Mg(OH)_2 \rightleftharpoons MgO+H_2O$ at $P_{H_2O}=P_{total}$. After Wyllie, p. 560.[55]

For the reaction: brucite $+ CO_2 \rightleftharpoons$ magnesite $+ H_2O$, $\nu_{H_2O} = -\nu_{CO_2}$, and each component occurs on opposite sides of the reaction. This makes ΔS very small, which makes $\partial T/\partial X_{CO_2}$ very large (Fig. 2.18). For the reaction: tremolite $+ 3$calcite $+ 2$quartz $\rightleftharpoons 5$diopside $+ 3CO_2 + H_2O$, we have:

$$\left(\frac{\partial T}{\partial X_{CO_2}}\right)_P = \frac{RT}{\Delta S}\left(\frac{3}{3}-\frac{1}{1}\right) = 0$$

so that the curve should lie parallel to the X-axis, which it does, but only at $X_{CO_2} = 0\cdot75$ (Fig. 2.18). At any other X value the curve has a TX slope, which is positive for $X_{CO_2} < 0\cdot75$ and negative for $X_{CO_2} > 0\cdot75$.

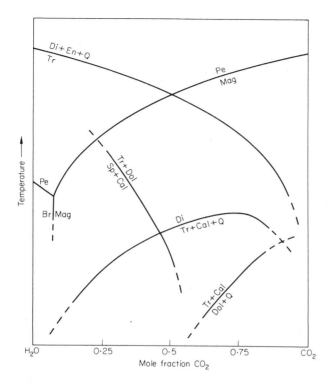

Fig. 2.18. Diagrammatic T-X_{CO_2} plot of various intersecting reaction curves, namely:

$Tr \rightleftharpoons 2Di + 3En + Q + H_2O$
$Mag \rightleftharpoons Pe + CO_2$
$Tr + 3Cal + 2Q \rightleftharpoons 5Di + 3CO_2 + H_2O$
   (T_{max} at $X_{CO_2} = 0.75$)
$Br + CO_2 \rightleftharpoons Mag + H_2O$
$5Dol + 8Q + H_2O \rightleftharpoons Tr + 3Cal + 7CO_2$
$4Sp + 9Cal + 5CO_2 \rightleftharpoons Tr + 7Dol + 7H_2O$

The positions of the curves are only approximate, and no attempt has been made to bring them into line with more recent experimental work, as they are meant only to make a general point (see text). After Greenwood, p. 84.[26]
Br=brucite; Cal=calcite; Di=diopside; Dol=dolomite; En=enstatite; Mag=magnesite; Pe=periclase; Q=quartz; Sp=serpentine; Tr=tremolite.

The possibility of *intersecting isograds* based on reactions involving varying proportions of H_2O and CO_2 as reactants and products is illustrated by figure 2.18. This suggests that

'isograds that are defined by reactions involving volatile components should not be regarded as isotherms even within the confines of a single map area, unless there is some evidence that the composition of the pore fluid was approximately constant over the area. This enhances, rather than reduces, the importance of mapping isograds. Displacement of isograds from their usual geographic relationships with other isograds, and sharp changes in trend of an isograd, can tell us more, in principle, about the composition of the pore fluids and pressure-temperature environment than isograds that appear in their "normal" positions.'[25]

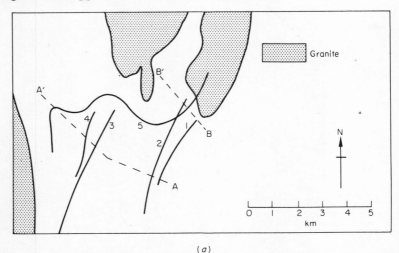

(*a*)

Fig. 2.19 (*a*)   Isograds in the Whetstone Lake area, Ontario, based on reactions:

(1) $Ch + Mu + Ga \rightleftharpoons Sta + Bi + Q + H_2O$
(2) $Ch + Mu + Sta + Q \rightleftharpoons Ky + Bi + H_2O$
(3) $Ky \rightleftharpoons Sill$
(4) $Sta + Mu + Q \rightleftharpoons Sill + Ga + Bi + H_2O$
(5) $Bi + Cal + Q \rightleftharpoons CaAmp + Kf + CO_2 + H_2O$

Note that the isograd representing reaction (5) intersects the other four. Amp = amphibole; Bi = biotite; Cal = calcite; Ch = chlorite; Ga = garnet; Kf = K-feldspar; Ky = kyanite; Mu = muscovite; Q = quartz; Sill = sillimanite; Sta = staurolite.

An excellent example of this occurs in the Whetstone Lake area, Ontario, where an isograd for a reaction producing $CO_2 + H_2O$ crosses several sub-parallel dehydration isograds in such a way as to suggest that the H_2O/CO_2 fugacity ratio was higher near a granite pluton (Fig. 2.19*a*).[9] The situation is shown on a TX diagram in Fig. 2.19*b*. An inferred polymorphic reaction (kyanite $\rightleftharpoons$ sillimanite) shows no T variation with changing X, because no fluid is involved. The dehydration reactions slope from a maximum temperature at $X_{H_2O} = 1$ ($X_{CO_2} = 0$) towards lower temperatures with decreasing X_{H_2O}. However, reaction (5) produces $H_2O + CO_2$ in the ratio of 1:3, so that its maximum temperature is for $X_{H_2O} = 0.25$ (where X has the composition of the fluid produced by the reaction), and the curve slopes to lower temperatures with both increasing and decreasing X_{H_2O}, as explained above. Therefore, two traverses (AA' and BB') in this area intersect different isograd sequences (Figs 2.19a, b).

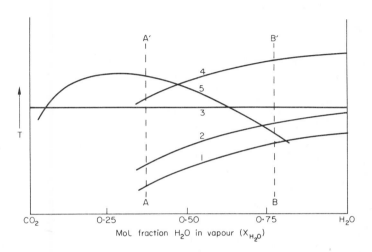

Fig. 2.19 (b)   T-X_{H_2O} diagram, for a constant P_{total} ($= P_{H_2O} + P_{CO_2}$) higher than that of the Al_2SiO_5 triple point showing schematically the equilibrium curves for reactions (1) to (5). Traverses AA' and BB' correspond to those shown on the map (a). After Carmichael, © 1970 Oxford University Press, by permission of The Clarendon Press, Oxford.[9]

Marked local variations in mineral assemblages within a metamorphic facies[52] or between different facies[23] may be caused by $\mu_{CO_2} : \mu_{H_2O}$ fluctuations on all scales, even on the scale of a single hand specimen (Chapter 1).

In rocks with an aqueous phase in the presence of graphite (a situation probably applicable to many metapelitic rocks[40]), the components in the fluid phase are C, O and H, but the molecular species may include CH_4, CO, CO_2, H_2O, H_2 and O_2, so that the fugacities of all these must be considered for a full understanding of reactions under these conditions.[19] The general tendency is for CH_4 to predominate at lower temperatures and for CO_2 and H_2O to predominate at higher temperatures, but the proportions of the various species vary considerably with temperature and the buffer assemblage used.[19] Relationships between gas fugacities in systems with a complex fluid phase, such as C-O-H and S-C-O-H, enable more complete understanding of geologically reasonable reactions involving three molecular species in a fluid phase, for example[19]:

$$4CaCO_3 + 2Fe_3O_4 + 6SiO_2 + H_2O + \tfrac{1}{2}O_2 \rightleftharpoons 2CaFe^{3+}Si_3O_{12}(OH) + 4CO_2$$
calcite   magnetite quartz   epidote

$$KFe_3AlSi_3O_{10}(OH)_2 + 3S_2 \rightleftharpoons KAlSi_3O_8 + 3FeS_2 + H_2O + {}^3/_2O_2$$
annite   K-feldspar   pyrite

Effect of Oxygen Fugacity[17, 19, 40]

Oxidation and reduction reactions are common in metamorphism, and so we must consider the fugacity of oxygen (or hydrogen), either in a fluid phase or in the solid parts of the rock. Differences in oxidation state between rocks and minerals can be detected chemically, but we have no reliable way of inferring the reactions responsible, except by experiment.

Expressions commonly used in general discussion are P_{O_2} (although the oxygen pressures in rocks are so small that they have no physical reality), f_{O_2}, μ_{O_2}, or a_{O_2}. For general and qualitative purposes only, they may be regarded as being approximately equivalent.

Because ferrous silicates are abundant, and iron oxide minerals have wide stability ranges, the following type of reaction should be common:

Fe-rich ferrous silicate + oxygen $\rightleftharpoons$ Fe-poor silicate + Fe oxide

Some examples are:

annite + oxygen $\rightleftharpoons$ K-feldspar + magnetite + water; and
hypersthene + oxygen + water $\rightleftharpoons$ talc + magnetite.

These reactions are controlled by f_{O_2} as well as pressure and temperature. They apply also to solid solutions, so the proportion of ferrous end-member in a mineral reflects the f_{O_2} at a given temperature.

Experimental work has established the relationship between f_{O_2} and temperature for various chemical equilibria involving oxides, ferrous silicates, and native metals in the presence of an O-H fluid (Fig. 2.20), so that observation of natural assemblages potentially can be used as a general guide for f_{O_2}-T conditions. For example, the occurrence of native Fe/Ni (awaruite) implies a relatively low f_{O_2}, although note that Ni is stable at higher f_{O_2} than Fe, so that the Ni stabilises awaruite to higher f_{O_2} than pure Fe. On the other hand, a higher f_{O_2} is indicated by the presence of magnetite or hematite. This work has been extended to include the effects of adding graphite, in the presence of a C-O-H fluid phase.[19]

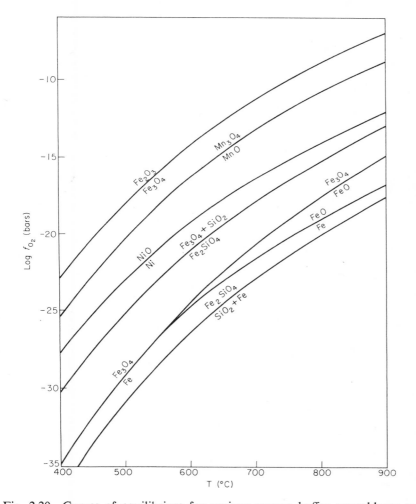

Fig. 2.20 Curves of equilibrium for various oxygen buffer assemblages as functions of oxygen fugacity (log f_{O_2}) and temperature for $P_{total} = 1$ atmosphere. Total pressure has little effect on these equilibria.
After Ernst (1968), *Amphiboles* in *Minerals, Rocks and Inorganic Materials*, Vol. 1, p. 35. Berlin-Heidelberg-New York: Springer-Verlag.

The f_{O_2} of a rock system may be controlled externally if oxygen is mobile, or internally if the system is closed to oxygen, and, as stated previously, the fugacities of the various molecular fluid species are variable, depending on temperature, solid buffer assemblage used and presence or absence of graphite.* For example, f_{H2} is reduced considerably in C-O-H systems, compared with O-H systems, using the same oxygen buffer assemblage.[19] So, a simple situation in natural rocks (especially those in which a C-O-H or S-C-O-H fluid may be suspected) cannot be assumed for devaporisation or redox reactions.

The oxygen content of layered metamorphic rocks generally appears to stay relatively constant during metamorphism. For example, in metamorphosed Pre-Cambrian iron-rich rocks of Michigan and Minnesota, thin magnetite-chert units are interlayered with hematite-chert units, suggesting that the different oxygen fugacities may be due to original sedimentary differences. However, where water was present in small amounts, reaction occurred across the rock unit boundaries, producing hematite in magnetite layers, and vice versa. Where more water was available, hydrous ferrous silicates (e.g. greenalite, minnesotaite and grunerite) formed in the magnetite-rich units only (where the f_{O2} was low enough to prevent their oxidation).[17]

Similarly, in the Glen Clova area, Scotland, the ratios of ferric to ferrous iron vary widely between intimately interlayered units in metapelitic gneiss, this being interpreted also as reflecting an original sedimentary difference.[10] Increasing oxidation of the rocks is accompanied by increasing muscovite and iron oxides and decreasing biotite and garnet, presumably reflecting oxidation of ferrous iron that otherwise would exist in the ferromagnesian silicates. The differences appear to be related to graphite content, in that graphite-bearing gneisses do not contain hematite. Furthermore, as would be expected, in more oxidised rocks the ferromagnesian minerals are more magnesium-rich, owing to removal of ion in oxide phases.

Coexisting Magnetite and Ilmenite

Experimental work on coexisting Fe-Ti oxides enables us to use the compositions of coexisting magnetite and ilmenite to estimate T and f_{O2} of metamorphism (Fig. 2.21), provided no post-metamorphic oxidation

*Nockleberg (1973, *Amer. J. Science*, **273**, 498–514) has calculated that at temperatures of 527–827°C and fluid pressures of 0·5-2kb, mixing of CO_2 with $H_2O + H_2$ fluids buffered to an f_{O2} within the graphite stability field causes extensive reaction of the fluids to form CO and CH_4, and oxidation of minerals buffering f_{O2} in the fluid. This may account for the high Fe^{3+}/Fe^{2+} values in some contact metamorphic calcsilicate rocks ('skarns') believed to have been formed by the action of aqueous fluids on carbonates.

or reduction of the oxide minerals has occurred.[6] The Ti content of magnetite is expressed as mol. per cent ulvöspinel (Fe_2TiO_4) and the excess Fe in ilmenite is expressed as mol. per cent hematite. Figure 2.21 shows that at constant T, increasing f_{O_2} causes increasing Fe^{3+} in both oxides; i.e. it produces more hematite in ilmenite and less ulvöspinel in magnetite.

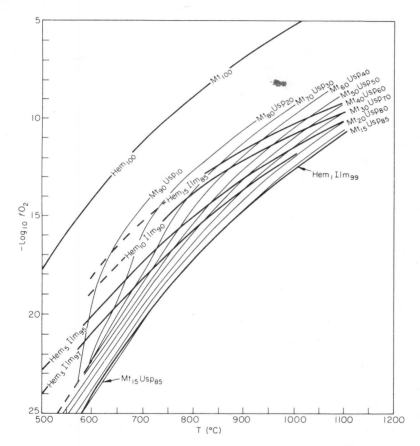

Fig. 2.21 T-f_{O_2} projection of experimentally determined magnetite-hematite equilibria, as affected by Ti solid solution, the compositions being expressed in mol per cent. The heavy dashed line at the bottom left represents temperature–composition relations of the magnetite-ulvöspinel solvus, although its f_{O_2} relationships are known only approximately. Mt = magnetite; Hem = hematite; Ilm = ilmenite; Usp = ulvöspinel.

After Buddington and Lindsley, p. 316, © 1964 Oxford University Press, by permission of The Clarendon Press, Oxford.[6]

Effect of f_{O_2} on Hydrous Silicates Without Cations of Variable Valency

It might be expected that f_{O_2} would have no effect on the stability of hydrous silicates without cations of variable valency. But Wones and Eugster have shown experimentally that low f_{O_2} in a fluid phase causes phlogopite to dehydrate at lower temperatures than at high f_{O_2}.[54] The reason is that, if O_2 is removed from the fluid (e.g. to enter an oxidised phase), f_{H_2O} must decrease in order to maintain the equilibrium constant of water:

$$K = (f_{H_2O})^2 / [(f_{H_2})^2 \times f_{O_2}]$$

assuming constant temperature. In other words, as f_{O_2} decreases, f_{H_2} must increase through dissociation of more water and consequent reduction in f_{H_2O}. This causes the dehydration reaction to run at a lower temperature, as discussed previously.

Distribution of Elements Between Coexisting Minerals

In the reactions discussed in this chapter I have not considered the distribution of elements between coexisting minerals. Because most common minerals are crystalline solutions of two or more components, the element distribution should be studied for the complete appreciation of metamorphic processes. Available chemical analyses suggest that the distribution of elements between stably coexisting minerals is orderly, especially in rocks of intermediate to high metamorphic grade. So, distribution coefficients may prove to be important indicators of equilibrium and of grade. The underlying principle is that an element generally tends to substitute preferentially into one of the two minerals concerned, equalising its chemical potential in both phases when the stable distribution has been achieved under the prevailing PTX conditions. Further information is provided by Saxena's new, comprehensive review of the thermodynamic theory of common crystalline solutions.[46]

An example of the possible use of distribution coefficients is provided by experimental work on the reaction: 3cordierite $\rightleftharpoons$ 2garnet + 4sillimanite + 5quartz, which is divariant in P-T space[31] (Fig. 2.2), as expected from the Phase Rule ($P = 4$, $C = SiO_2 + Al_2O_3 + MgO + FeO = 4$, $\therefore F = 2$). Magnesium tends to be strongly partitioned into cordierite and iron into garnet, so that within the divariant field (Fig. 2.2), $X_{Cd} > X_{Ga}$ (where $X = Mg/Mg + Fe^{2+}$). So the initial tendency is for an iron-rich garnet to be produced, after which increasing pressure causes a gradual increase in the garnet:cordierite ratio, accompanied by a gradually increasing magnesium content in the garnet and cordierite. This means that, in principle, the X_{Cd} and X_{Ga} of natural garnet-cordierite assemblages can position them in the experimentally determined divariant field (Fig. 2.2), although the presence of water and iron oxides in natural rocks could complicate matters. This approach could be important for many similar sliding reactions.

References

1 Anderson, G. M. (1970). Some thermodynamics of dehydration equilibria. *Amer. J. Science*, **269**, 392–401.
2 Atherton, M. P. (1965). The chemical significance of isograds. In *Controls of Metamorphism*, ed. Pitcher, W. S. and Flinn, G. W. 169–202. Edinburgh: Oliver & Boyd.
3 Bird, G. W. and Anderson, G. M. (1973). The free energy of formation of magnesian cordierite and phlogopite. *Amer. J. Science*, **273**, 84–91.
4 Blackburn, W. H. (1968). The spatial extent of chemical equilibrium in some high-grade metamorphic rocks from the Grenville of southeastern Ontario. *Contribs. Mineralogy & Petrology*, **19**, 72–92.
5 Brown, E. H. (1971). Phase relations of biotite and stilpnomelane in the greenschist facies. *Contribs. Mineralogy & Petrology*, **31**, 275–99.
6 Buddington, A. F. and Lindsley, D. H. (1964). Iron-titanium oxide minerals and synthetic equivalents. *J. Petrology*, **5**, 310–57.
7 Burnham, C. W., Holloway, J. R. and Davis, N. F. (1969). Thermodynamic properties of water to 1,000°C and 10,000 bars. *Geol. Soc. America Special Paper* no. 132.
8 Carmichael, D. M. (1969). On the mechanism of prograde metamorphic reactions in quartz-bearing pelitic rocks. *Contribs. Mineralogy & Petrology*, **20**, 244–67.
9 Carmichael, D. M. (1970). Intersecting isograds in the Whetstone Lake area, Ontario. *J. Petrology*, **11**, 147–81.
10 Chinner, G. A. (1960). Pelitic gneisses with varying ferrous/ferric ratios from Glen Clova, Angus, Scotland. *J. Petrology*, **1**, 178–217.
11 Chinner, G. A. (1962). Almandine in thermal aureoles. *J. Petrology*, **3**, 316–40.
12 Coe, R. S. and Paterson, M. S. (1969). The α-β inversion in quartz: a coherent phase transition under nonhydrostatic stress. *J. Geophysical Research*, **74**, 4921–48.
13 Day, H. W. (1972). Geometrical analysis of phase equilibria in ternary systems of six phases. *Amer. J. Science*, **272**, 711–34.
14 Denbigh, K. (1966). *The Principles of Chemical Thermodynamics*. 2nd edn. London: Cambridge University Press.
15 Epstein, S. and Taylor, H. P. (1967). Variation of O^{18}/O^{16} in minerals and rocks. In *Researches in Geochemistry*, ed. Abelson, P. H. vol. 2, 29–62. New York: J. Wiley & Sons, Inc.
16 Ernst, W. G. (1968). *Amphiboles*. Ch. IV. Hydrothermal phase equilibration and natural stability, 33–9. Berlin-Heidelberg-New York: Springer-Verlag.
17 Eugster, H. P. (1959). Reduction and oxidation in metamorphism. In *Researches in Geochemistry*, ed. Abelson, P. H. vol. 1, 397–426. New York: J. Wiley & Sons, Inc.

18 Eugster, H. P. (1970). Thermal and ionic equilibria among muscovite, K-feldspar and aluminosilicate assemblages. *Fortschr. Miner.*, **47**, 106–23.

19 Eugster, H. P. and Skippen, G. B. (1967): Igneous and metamorphic reactions involving gas equilibria. In *Researches in Geochemistry*, ed. Abelson, P. H. vol. 2, 492–520. New York: J. Wiley & Sons, Inc.

20 Fisher, J. R. and Zen, E. (1971). Thermochemical calculations from hydrothermal phase equilibrium data and the free energy of formation of H_2O. *Amer. J. Science*, **270**, 297–314.

21 Fry, N. and Fyfe, W. S. (1969). Eclogites and water pressure. *Contribs. Mineralogy & Petrology*, **24**, 1–6.

22 Fyfe, W. S., Turner, F. J. and Verhoogen, J. (1958). Metamorphic reactions and metamorphic facies. *Mem. Geol. Soc. America*, no. 73.

23 Glassley, W. (1974). A model for phase equilibria in the prehnite-pumpellyite facies. *Contribs. Mineralogy & Petrology*, **43**, 317–32.

24 Gordon, T. M. (1973). Determination of internally consistent thermodynamic data from phase equilibrium experiments. *J. Geology*, **81**, 199–208.

25 Greenwood, H. J. (1961). The system $NaAlSi_2O_6$-H_2O-argon: total pressure and water pressure in metamorphism. *J. Geophys. Research*, **66**, 3923–46.

26 Greenwood, H. J. (1962). Metamorphic reactions involving two volatile components. *Ann. Report Director Geophysical Laboratory, Carnegie Inst. Washington Year Book*, **61**, 82–5.

27 Greenwood, H. J. (1967). Mineral equilibria in the system MgO-SiO_2-H_2O-CO_2, In *Researches in Geochemistry*, ed. Abelson, P. H. vol. 2, 542–67. New York: J. Wiley & Sons, Inc.

28 Helgeson, H. C. (1967). Solution chemistry and metamorphism, In *Researches in Geochemistry*, ed. Abelson, P. H. vol. 2, 362–404. New York: J. Wiley & Sons, Inc.

29 Helgeson, H. C. (1969). Thermodynamics of hydrothermal systems at elevated temperatures and pressures. *Amer. J. Science*, **267**, 729–804.

30 Hemley, J. J. (1959). Some mineralogical equilibria in the system K_2O-Al_2O_3-SiO_2-H_2O. *Amer. J. Science*, **257**, 241–70.

31 Hensen, B. J. (1971). Theoretical phase relations involving cordierite and garnet in the system MgO-FeO-Al_2O_3-SiO_2. *Contribs. Mineralogy & Petrology*, **33**, 191–214.

32 Hollister, L. S. (1969). Contact metamorphism in the Kwoiek area of British Columbia: an end member of the metamorphic process. *Bull. Geol. Soc. America*, **80**, 2465–94.

33 Kerrick, D. M. (1972). Experimental determination of muscovite + quartz stability with $P_{H2O} < P_{total}$. *Amer. J. Science*, **272**, 946–58.

34 Korzhinsky, D. S. (1950). Phase rule and geochemical mobility of elements. *International Geological Congress, Rept.* 18th Session, Part II, 50–7.

35 Korzhinsky, D. S. (1950). Differential mobility of components and metasomatic zoning in metamorphism. *International Geological Congress, Rept.* 18th Session, Part II, 65–72.

36 Korzhinsky, D. S. (1959). *Physicochemical Basis of the Analysis of the Paragenesis of Minerals.* New York: Consultants Bureau.

37 Krauskopf, K. B. (1967). *Introduction to Geochemistry*, New York: McGraw-Hill Book Co.

38 Kretz, R. (1966). Interpretation of the shape of mineral grains in metamorphic rocks. *J. Petrology*, **7**, 68–94.

39 Kwak, T. A. P. (1974). Natural staurolite breakdown reactions at moderate to high pressures. *Contribs. Mineralogy & Petrology*, **44**, 57–80.
40 Miyashiro, A. (1964). Oxidation and reduction in the Earth's crust with special reference to the role of graphite. *Geochimica et Cosmochimica Acta*, **28**, 717–29.
41 Paterson, M. S. (1973). Nonhydrostatic thermodynamics and its geologic applications. *Reviews of Geophysics & Space Physics*, **11**, 355–89.
42 Raleigh, C. B. and Paterson, M. S. (1965). Experimental deformation of serpentinite and its tectonic implications. *J. Geophys. Research*, **70**, 3965–85.
43 Robie, R. A. and Waldbaum, D. R. (1968). Thermodynamic properties of minerals and related substances at 298·15°K (25·0°C) and one atmosphere (1·013 bars) pressure and at higher temperature. *U.S. Geol. Survey Bull.*, no. 1259.
44 Rosenfeld, J. L. (1961). The contamination-reaction rules. *Amer. J. Science*, **259**, 1–23.
45 Ross, J., Papike J. J. and Shaw, K. W. (1969). Exsolution textures in amphiboles as indicators of subsolidus thermal histories. *Mineral. Soc. Amer. Special Paper* no. 2, 275–99.
46 Saxena, S. K. (1973). *Thermodynamics of Rock-forming Crystalline Solutions.* Berlin-Heidelberg-New York: Springer-Verlag.
47 Thompson, J. B. (1955). The thermodynamic basis for the mineral facies concept. *Amer. J. Science*, **253**, 65–103.
48 Thompson, J. B. (1959). Local equilibrium in metasomatic processes. In *Researches in Geochemistry*, ed. Abelson, P. H. vol. 1, 427–57. New York: J. Wiley & Sons, Inc.
49 Turner, F. J. (1968). *Metamorphic Petrology. Mineralogical and Field Aspects.* New York: McGraw-Hill Book Co.
50 Vidale, R. J. and Hewitt, D. A. (1973). 'Mobile' components in the formation of calc-silicate bands. *Amer. Mineralogist*, **58**, 991–7.
51 Wall, V. J. and Essene, E. J. (1972). Subsolidus equilibria in CaO-Al_2O_3-SiO_2-H_2O. *Geol. Soc. America Abstracts*, **4**, 700.
52 Watts, B. J. (1973). Relationship between fluid-bearing and fluid-absent invariant points and a petrogenetic grid for a greenschist facies assemblage in the system CaO-MgO-Al_2O_3-SiO_2-CO_2-H_2O. *Contribs. Mineralogy & Petrology*, **40**, 225–38.
53 Winkler, H. G. F. (1967). *Petrogenesis of Metamorphic Rocks* 2nd edn, Berlin-Heidelberg-New York: Springer-Verlag.
54 Wones, D. R. and Eugster, H. P. (1965). Stability of biotite: experiment, theory, and application. *Amer. Mineralogist*, **50**, 1228–72.
55 Wyllie, P. J. (1962). The petrogenetic model, an extension of Bowen's petrogenetic grid. *Geol. Mag.*, **99**, 558–69.
56 Zen, E. (1963). Components, phases, and criteria of chemical equilibrium in rocks. *Amer. J. Science*, **261**, 929–42.
57 Zen, E. (1966). Construction of pressure-temperature diagrams for multicomponent systems after the method of Schreinemakers—a geometrical approach. *U.S. Geol. Survey Bull*, no. 1225.
58 Zen, E. (1966). Some topological relationships in multisystems of n+3 phases. I. General theory; unary and binary systems. *Amer. J. Science*, **264**, 401–27.
59 Zen, E. (1967). Some topological relationships in multisystems of n+3

phases. II. Unary and binary metastable sequences. *Amer. J. Science*, **265**, 871–97.

60 Zen, E. (1969). Free energy of formation of pyrophyllite from hydro-thermal data: values, discrepancies and implications. *Amer. Mineralogist*, **54**, 1592–1606.

61 Zen, E. and Roseboom, E. H. (1972). Some topological relationships in multisystems of n+3 phases. III. Ternary systems. *Amer. J. Science*, **272**, 711–34.

Chapter 3

Kinetics of
Metamorphic Reactions

Introduction

Metamorphic rocks exposed at the earth's surface are *metastable*, with respect to P-T-X conditions in the atmosphere. This is because of the extremely slow rate of reaction between the rocks and the atmosphere, and between the minerals of the rocks themselves, under these low-temperature conditions. If this were not so, the rocks would contain only low P-T assemblages, and we would have no evidence of metamorphism ever having occurred at depth.

Many metamorphic rocks show microstructural evidence of incomplete reactions (especially retrograde reactions), which could be due partly to (i) restricted availability of reactants (such as water in many retrograde reactions), or (ii) slow reaction rates, even if reactants are available (i.e. kinetic difficulties). We also need to appreciate the kinetics of metamorphic reactions in order to evaluate the results of experimentalists. For example, we must ask whether or not their results represent a close approach to stability of phases. Of course, it must be appreciated that there is no way of ensuring that a stable (lowest possible free energy) assemblage has been produced, either in a natural rock or an experiment. But at least we should try to find out what is the most stable (even if metastable) assemblage we can obtain under a given set of conditions, by eliminating or evaluating as many kinetic factors as possible.

All thermally activated processes (such as diffusion, recrystallisation and chemical reactions) are opposed by free energy barriers, which must be overcome before the processes can take place at appreciable rates, even if they are thermodynamically favourable (Fig. 3.3).[3,7] For example, in metamorphic reactions, *nucleation* of new phases must occur before *growth* can take place. Nucleation is opposed by energy barriers (discussed later), as is growth. So we must examine the nature and effectiveness of these barriers.[7]

Experimental determinations of the variation in reaction rate with temperature typically give similar results that can be expressed in the general (Arrhenius) equation:[3]

$$\text{reaction rate} = \text{constant} \times e^{-Q/RT}$$

where Q = activation energy (calories/mole), R = gas constant (=1·986 cal/mole/°K), T = absolute temperature (°K), and the constant is independent of temperature. Broadly speaking, this rate constant (also called the 'frequency factor') expresses the number of times per second that a reacting particle enters into the reaction, whereas $e^{-Q/RT}$ expresses the fraction of all reacting particles that have an energy (Q) above the average energy of all particles present. Conventionally, logs to the base 10 of the Arrhenius equation are taken, giving:

$$\log_{10} \text{rate} = \log_{10} \text{constant} - [(Q/2\cdot303R) \times (1/T)]$$

which has the algebraic form of a straight line ($y = a + bx$), so that the slope is $-Q/2\cdot303R$ and the intercept on the y-axis is $\log_{10}$ constant. So, by plotting the experimentally determined reaction rate against temperature (Fig. 3.1), a value for the activation energy (Q) and for the rate constant can be obtained.[3]

If Q is large the slope is steep, and vice versa (Fig. 3.2). Note that the dependence of rate on T is logarithmic, so if the slope is steep the reaction rate will increase very rapidly with increasing temperature, especially at higher temperatures. Q reflects the dependence of rate on temperature; for example, if Q is small, we can say that the reaction rate is not very temperature-dependent. But the rate also depends on the rate constant, which can outweigh the effect of Q (Fig. 3.2*c*).

The activation energy expresses the free energy barrier referred to previously (Fig. 3.3). If a number of alternative paths for a reaction are possible, the one with the lowest free energy barrier will be favoured— possibly a path involving the catalytic effect of another phase (Fig. 3.3). Commonly, a step process (i.e. a series of sub-reactions) is kinetically more favourable than a single reaction process (Fig. 3.3*c*). In fact, this is quite common in chemical reactions (Ostwald's step rule).[7] In such a series of reactions the overall rate of the net reaction is determined by the rate of the slowest step.

Unfortunately, experimental studies of the rates of metamorphic silicate reactions are so time-consuming and expensive that too few

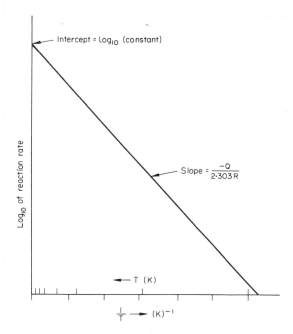

Fig. 3.1 Typical Arrhenius plot of experimental data fitting the equation $\log_{10}$ rate $= \log_{10}$ (constant) $- (Q/2 \cdot 303R) \times (1/T)$. Note the logarithmic scale for T and the arithmetic scale for $1/T$. After Brophy, Rose and Wulff, p. 64.[3]

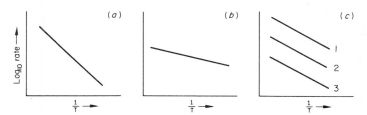

Fig. 3.2 Hypothetical Arrhenius plots. (*a*) High Q (steep slope); (*b*) Low Q (shallow slope); (*c*) Arrhenius lines for three reactions with same Q but different frequency factors so that the relative reaction rates are $1 > 2 > 3$.

have been carried out.[10, 14] A notable example is the work of Greenwood who showed that, though anthophyllite initially grows rapidly from talc in the stability field of quartz + enstatite, it eventually breaks down if the experiments are allowed to run for long enough times (Fig. 3.4).[10] So the reaction: talc $\rightleftharpoons$ enstatite + quartz + vapour prefers to take place by steps involving the intermediate production of anthophyllite. Short-time experiments (up to about 1 500 minutes) would give the false im-

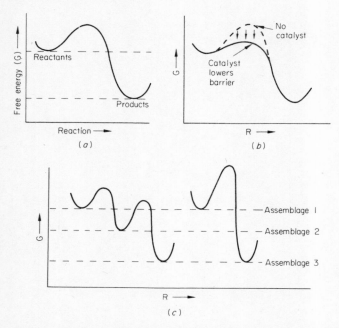

Fig. 3.3   (*a*) Free energy barrier between reactants and products; (*b*) lowering of the energy barrier by the presence of a catalyst; (*c*) energy barriers between three different assemblages, such that the rate of transition $1 \rightarrow 2 > 2 \rightarrow 3 > 1 \rightarrow 3$.

pression that anthophyllite is the stable phase at 830°C and 1kb in this system. We must be aware of this kind of kinetic problem when applying experimental results to natural assemblages.

One of the main factors controlling the nucleation and growth of metamorphic minerals is *diffusion*, itself a process whose rate is temperature-dependent.

Diffusion in Metamorphism[3, 6, 7, 12, 13, 20]

Consider a simple reaction: $A + B \rightleftharpoons C + D$. An increase in the concentration of C or D will slow down or even stop the reaction. In order for the reaction to proceed, (1) products must escape from the reaction site, and (2) reactants (or their components) must remain mutually accessible. This emphasises the importance of diffusion in metamorphic reactions. If, for example, a layer of products (C+D) were built up between grains of reactants (A+B), the reaction would stop once the components of A and B could not diffuse fast enough through the boundary layer.

Similarly, in devaporisation reactions (e.g. $A+B \rightleftharpoons C+D+V$) the volatile phase (V) must escape from the reaction site if the reaction is to proceed.

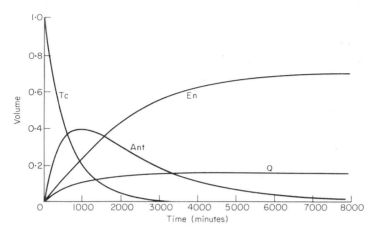

Fig. 3.4   Volumes of phases produced during the heating of talc (Tc) at 830°C and 1 kb with changing time. Note that anthophyllite (Ant) initially is produced metastably, but that, in time, enstatite (En) and quartz (Q) increase as anthophyllite decreases. Thus, the stable assemblage under these P-T conditions is enstatite + quartz, not anthophyllite, as short-time experiments would indicate. After H. J. Greenwood (1963). *J. Petrology*, **4**, pp. 317–51. © 1963 Oxford University Press, by permission of The Clarendon Press, Oxford.

Diffusion is the process by which atoms, molecules or ions move from one position to another within a solvent phase, under the influence of a chemical potential gradient.[12] In solids this means the periodic jumping of atoms from one site in the structure to another. Obviously such jumps are obstructed by neighbouring atoms. Their rate is controlled strongly by temperature, which gives atoms the necessary energy to surmount the energy barrier (Fig. 3.3) opposing their thermal vibration beyond the limits of their normal positions. The jumps are random, but, if a *chemical potential gradient* (e.g. a pressure, temperature or concentration gradient) is operating, there will be a greater tendency for more atoms to go one way (down the gradient) than the other. The net result is displacement of matter.

Self diffusion is diffusion in a pure substance, which can be detected best by radioactive tracer atoms added to the substance. *Interdiffusion* is diffusion of one component in the lattice of another. Diffusion is isotropic for cubic solids and anisotropic for others.

Diffusion can be treated macroscopically, by assuming the material is a continuum and ignoring the atomic structure.[3, 7, 12] The mathematical analysis is over a hundred years old, but is still valid for the macroscopic handling of experimental data. Consider the situation in figure 3.5. The flux (J) is defined as the amount of material passing through unit area normal to the flux direction per unit time. The initial concentration (C_A) and final concentration (C_B) are constant, but, since $C_A > C_B$, the concentration gradient is negative from left to right. The flux increases with an increasingly negative gradient (dC/dx). The coefficient of proportionality in this relationship between flux and gradient is called the *diffusion coefficient* (diffusivity), **D**:

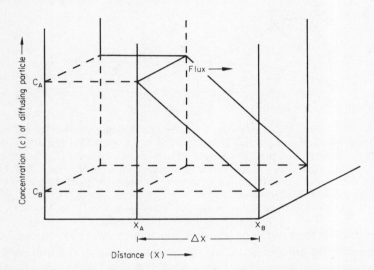

Fig. 3.5   Simplest possible diffusion system, in which the flow (flux) of diffusing particles is positive from left to right, as the particles move from an initial high concentration (C_A) at X_A to a lower concentration (C_B) at X_B over a distance ΔX, under a steady state gradient.

$$J = -D\frac{dC}{dx} \quad \text{(Fick's First Law)}$$

This is for a steady state situation, but more commonly the concentration of diffusing material changes with time, so that dC/dx varies with time (t):

$$\frac{dC}{dt} = \frac{d}{dx}\left(D\frac{dC}{dx}\right) \quad \text{(Fick's Second Law)}$$

Experimentally, D is obtained from the relationship $X=(Dt)^2$, where X=distance of transfer of the average concentration of the diffusing material, and t=time. The experiments are repeated at different temperatures (T), so that $\log_{10}D$ can be plotted against $1/T$, obtaining the usual Arrhenius plot for thermally activated processes, given by $D=D_O e^{-Q/RT}$. Then D_O and Q can be read from the graph.

Diffusion is connected with the presence of *point defects*[6, 13, 17] (such as vacancies, substitutional impurity atoms and interstitial impurity atoms) in all crystalline materials. In ionic solids, electrical neutrality has to be maintained, and the most common imperfections are Schottky and Frenkel defects (Fig. 3.6). The main diffusion mechanisms are (i) *motion of vacancies* (thereby causing a motion of diffusing particles in the opposite direction) and (ii) *motion of small interstitial cations*. Because restraining atoms must be distorted (moved apart far enough), an energy barrier must be surmounted before diffusion will occur. The activation energy (Q) for diffusion can be lowered by increasing the concentration of impurity atoms and other defects (Fig. 3.7), including dislocations produced by deformation.[13]

Diffusion is also faster along grain boundaries and surfaces (against the atmosphere or another fluid) than through the volume of a crystalline material. The following relationships apply:[3]

$$Q_{\text{volume diffusion}} > Q_{\text{grain boundary diffusion}} > Q_{\text{surface diffusion}}$$

$$D_{\text{surface diffusion}} > D_{\text{grain boundary diffusion}} > D_{\text{volume diffusion}}$$

However, these differences become smaller at high temperatures, so that, for example, the kinetics of volume and grain boundary diffusion in alkali feldspar probably are similar at about 1 000°C.[16]

Much more experimental work on diffusion in silicates is needed. For self-diffusion of K^+ and Na^+ in alkali feldspar, available experiments (using radioactive tracers) have given the following approximate results:[16, 19]

K+		Na+	
D (cm² sec⁻¹)	T (°C)	D (cm² sec⁻¹)	T (°C)
10^{-11}	1 050	10^{-9}	1 000–1 050
10^{-12}	800–1 000	10^{-10}	850–1 000
10^{-13}	800–900	10^{-11}	700–800
10^{-14}	700–850	10^{-12}	600–700
$10^{-15.5}$	600	10^{-13}	500

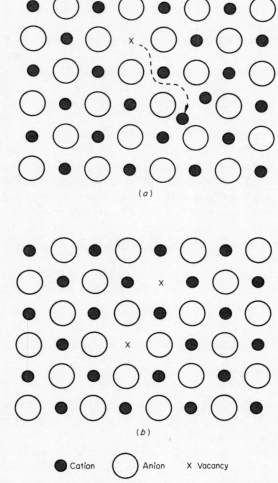

Fig. 3.6   Imperfections in ionic crystals. (*a*) Frenkel disorder, in which a cation has left a normal site (surmounting an energy barrier to do so) for an interstitial position, leaving a vacancy; (*b*) Schottky defect, in which equal numbers of cation and anion vacancies are present. After Kingery, pp. 167, 171.[13]

However, for each cation, the diffusion coefficient increases with increasing K/(Na+K) ratio in the alkali feldspar. The results show that volume diffusion of Na^+ is faster than K^+ in alkali feldspar for a given temperature. However, the activation energies are similar for diffusion of each ion, being about 60 ± 10 kc/mole.[19] Since the Si-O bond energy is about 88 kc/mole, the diffusion cannot involve breaking of Si-O bonds without a catalyst. Hydroxyl would be the most probable natural catalyst, but some experiments indicate that P_{H_2O} has no appreciable effect on the diffusion coefficients concerned.[19] The same reasoning suggests that Frenkel defects, rather than Schottky defects, control the diffusion. Furthermore, consideration of the crystal structure of alkali

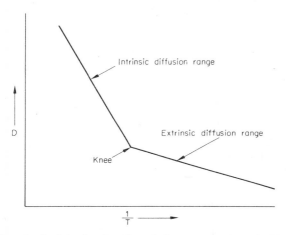

Fig. 3.7 Hypothetical Arrhenius plot of the type common in NaCl and some oxide crystals. The high-temperature part (intrinsic diffusion range) has a larger Q and is the same for many different specimens of the same material, regardless of their history; D is as expected for an equilibrium concentration of Schottky and Frenkel defects. The low-temperature part (extrinsic diffusion range) has a smaller Q owing to the presence of impurities and the effect of history (e.g. the introduction of dislocations during deformation). The 'knee' occurs at different places on the line for different specimens of the same material (depending on their purity and stored defects).

feldspar suggests that though interstitial diffusion in the [001] direction is possible, a vacancy mechanism is much more probable.[19]

Limited experimental work[19] indicates that although alkali diffusion in alkali feldspar is approximately isotropic in the (010) plane, $D_{[110]}$ is about a hundred times larger than $D_{[010]}$ at 890–1 000°C. Anisotropy of diffusion always should be kept in mind as a possibility for silicates.

Calculations[7] for quartz (based on compressibility data) suggest that a Q of 20 kc/mole would be involved in the crowding apart of lattice ions to make a hole of 1 Å. A Na^+ ion (0·98 Å) could do this, but for a K^+ ion (1·33 Å), Q would have to be 47 kc/mole. Since the calculated Q for diffusion of Na^+ and K^+ in quartz is 24 and 27 kc/mole, respectively, the conclusion is that Na^+ conceivably may diffuse in quartz by an interstitial mechanism, but not K^+, which presumably employs a vacancy mechanism.

The temperatures necessary to achieve a value for D of 10^{-10} for Na^+ in various sodium minerals[21] vary from 260°C (for the open-structured analcite) to over 950°C (for acmite). Diffusion of Na^+ in obsidian (disordered structure) is relatively rapid, and progressively decreases for cryolite (ionic), felspathoids and feldspars. The value for D of 10^{-10} is believed to be necessary for effectiveness in reactions.

An approximate diffusion coefficient for Mg/Fe exchange in garnet at 505°C is 10^{-15} cm²/sec, as indicated by available experimental data.[1] Extrapolation of experimental information on diffusion in Al_2O_3 suggests diffusion coefficients of only about 10^{-16} cm²/sec at 1 200°C for Al in other oxides and silicates.[1] The same reasoning applies to Si, so that diffusion of Si and Al, compared with other cations, should be negligible at metamorphic temperatures. Si and Al typically are distributed homogeneously in garnet zoned with respect to Mg, Fe, Mn, and Ca.[1]

Since oxygen constitutes almost all the volume of oxides and silicates, the net flux of oxygen must be negligible if the volume of the grain or rock volume under discussion is held constant.[1] However, this is not to say that exchange of oxygen atoms cannot occur during metamorphism. Yund (personal communication, 1973) has found that the activation energy for diffusion of oxygen in K-feldspar is much smaller than for K^+ in the same K-feldspar under identical conditions (Fig. 3.8), and also that the rate of oxygen diffusion is several orders of magnitude greater than K^+ diffusion at relatively low temperature (say, 500°C). Probably oxygen diffusion is assisted by hydrolysis of O-Si bonds in the feldspar by the abundant water present in the experiments, but the exact diffusion mechanism is unknown. However, the experiments demonstrate dramatically the rapidity of oxygen exchange in silicates in the presence of water.

Experiments at 700°C and 2 kb show that, for oxygen diffusion in a natural phlogopite, $D = 3 \times 10^{-16}$ cm²/sec and Q is about 30 kc/gm atom.[8] Diffusion of tetrahedrally co-ordinated oxygen is involved, and appears to take place mainly parallel to the c-axis. The diffusion coefficient for oxygen under these conditions is similar to that for K, but is only 10^{-3} that of Na. Diffusion of K in mica may be largely normal to the c-axis.

Nucleation and Growth[3, 7, 13]

Phase changes do not occur by the co-operative action of every atom in the assemblage, but are initiated by local fluctuations in energy that permit kinetic barriers to be surmounted. In fact, nuclei (only a few hundred atoms in size) of the stable phase(s) are formed. Some of these nuclei may disappear, whereas others grow to form grains. The various steps involved are as follows:

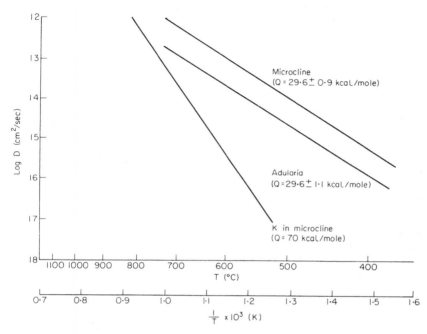

Fig. 3.8 Plot of D vs. $1/T$ for diffusion of oxygen in K-feldspar using O^{18} tracer for $P_{H_2O} = 2$ kb and in the presence of 2 molar KCl solution. Note the identical activation energies for diffusion in microcline and adularia, and the much larger activation energy for diffusion of K^+ in the same microcline under identical conditions, as discussed in text. After R. A. Yund (work in preparation).

Nucleation   (a) Assembly of the right atoms by diffusion;
(b) change of the atomic arrangement into one or more unstable intermediate structures ('activated complexes');
(c) formation of nuclei of a new stable (or metastable) phase.

Growth   (a) Transfer of the required material by diffusion through the old phase;
(b) transfer across the phase boundary into the new growing phase;

(c) transfer into the interior of the new phase;

(d) removal of unwanted material from the reaction site.

All these steps are thermally activated, and the one with the highest Q will control the Q for the overall process.

Nucleation   *Homogeneous nucleation* is the random formation of a nucleus in a homogeneous material, such as a melt or a perfect crystal. The production of a volume of a new phase causes a decrease in free energy of the system, but a new interface (between new and old phase) is formed also, and this involves an increase in free energy. So, the situation is:

$$\Delta G_{reaction} = -\Delta G_{volume} + \Delta G_{interface}$$
$$= -4/3\pi r^3 \Delta G_v + 4\pi r^2 \gamma$$

Because the nuclei are small, the surface/volume ratio is high, so that the interfacial free energy (γ) tends to dominate until a certain critical radius (r_c) is achieved, after which the volume term takes over. Embryos have $r < r_c$ and tend to re-dissolve; critical clusters have $r = r_c$; and nuclei have $r > r_c$ and tend to grow. Once again, an increment of energy (ΔG_c) has to be supplied to an embryo to enable it to surmount the energy barrier and become a stable nucleus. This causes overstepping during reactions (i.e. a temperature-dependent reaction tends to run at an appreciable rate only at some temperature above or below the thermodynamic equilibrium temperature).

Nucleation rates in solids are reduced by slower diffusion rates and an increased elastic strain caused by the 'room problem' of new phases growing in other solid phases for reactions with an appreciable ΔV. This tendency towards relatively sluggish nucleation means that these reactions proceed at very small rates near the equilibrium curve, and require more overstepping than (for example) crystallisation in liquids.

Heterogeneous nucleation is assisted by a 'nucleating agent', which lowers the free energy barrier, compared with homogeneous nucleation. For example, some solid material generally is present, even in liquid → solid transformations (e.g. in magmas), so that the solid could act as a nucleating agent ('substrate') as explained in figure 3.9. The balance of interfacial energies ('tensions') is:

$$\gamma \, SL \cos \theta + \gamma \, NS = \gamma \, NL$$

The NS interface replaces an equivalent amount of NL interface, and, since $\gamma \, NS < \gamma \, NL$, the result is a decrease in total

γ, although the shape of the nucleus is not spherical. Therefore, $\Delta G_{c \text{ (heterogeneous)}} < \Delta G_{c \text{ (homogeneous)}}$, which can make nucleation easier and reduce the amount of overstepping necessary for reactions to occur. So crystals tend to nucleate on other crystals.

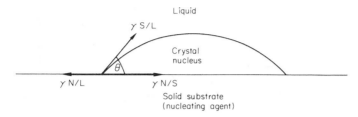

Fig. 3.9 System of interfacial energies ('tensions') operating for a non-spherical nucleus that is nucleating heterogeneously on a solid substrate. The contact angle is θ. S=solid (crystal), L=liquid, N=nucleating agent.

We can expect that most reactions nucleate heterogeneously, especially in solids, where (i) ΔV causes strains, (ii) various defects are present with their own strain effects, which may be reduced by nucleation of a new phase at their sites, (iii) diffusion occurs more readily in the presence of defects and (iv) the existence of a crystal structure may enable *epitaxial* nucleation (i.e. use of the old atomic structure by the new phase). Epitaxis might cause growth of metastable phases, if their structure is more closely related to that of the old phase than is the structure of the thermodynamically stable phase(s). Epitaxial nucleation and growth occur in some common metamorphic situations; e.g. oriented sillimanite intergrowths in biotite, and the parallel fibres of amphibole ('uralite') replacing clinopyroxene.

In crystalline solids, nucleation commonly occurs along grain boundaries or deformation structures (e.g. kink bands). As a nucleus is formed, an area of grain boundary is removed, thereby providing an energy source that can reduce the amount of energy needed from thermal fluctuation. For nucleation of ferrite (α-iron) in a polygrain matrix of austenite (γ-iron), junctions of four grains are favoured nucleation sites, followed by three- and two-grain junctions (Table 3.1).

Table 3.1: *Maximum energy for formation of ferrite nucleus in polygrain austenite (calculated from known austenite/austenite and ferrite/austenite grain boundary energies).*[5]

Grain interior	$1 \cdot 913 \times 10^{-8}$ ergs
2-grain junction	$0 \cdot 220 \times 10^{-8}$ ergs
3-grain junction	$0 \cdot 0436 \times 10^{-8}$ ergs
4-grain junction	$0 \cdot 0096 \times 10^{-8}$ ergs

Nucleation is assisted greatly by lattice strains (i.e. density of dislocations; Chapter 6) produced during deformation, especially in polymorphic transformations. The stored strain energy raises the free energy of the plastically deformed polymorph, which may permit metastable nucleation and growth of another polymorph outside its thermodynamic stability field. For example, coesite (a high-pressure polymorph of SiO_2) has been grown experimentally in strongly deformed quartz at confining pressures well below the quartz $\rightleftharpoons$ coesite equilibrium transformation pressure.[9,11] Similarly, aragonite has been grown experimentally in strained calcite at pressures below the aragonite $\rightleftharpoons$ calcite equilibrium.[18] The geological implications are discussed later in this chapter.

Growth   Growth is possible once ΔG_2 is exceeded. Material must be transferred through the phases involved, so that volume diffusion (with time-dependent concentrations) may control the overall growth rate. Therefore, the grain-boundary movement of a growing phase is time-dependent. Because nucleation and growth go on at the same time, the overall reaction rate is proportional to some product of $\dot{N}$ and $\dot{G}$ (where $\dot{N}$=nucleation rate and $\dot{G}$=growth rate).

Application of Reaction Kinetics to Metamorphism

Partly completed retrograde reactions (especially where pseudomorphous or vein-like) are readily recognisable, but the problem is how to be sure of equilibrium assemblages in the absence of such obvious criteria (Chapter 2). The uniformity of assemblages from place to place could be a universal attainment of metastability, but this need not preclude attempts to relate assemblages to P-T-X conditions, as long as the same degree of metastability were attained everywhere. The metamorphic facies scheme (Chapter 1), as defined originally, may have assumed general equilibrium, but this is not essential for its successful application.[22]

As mentioned previously, most metamorphic assemblages represent a peak of metamorphism, and only locally revert to lower grade assemblages on cooling, except in zones of later deformation, possible re-heating and supply of volatile material.

The reason may be that reactions proceed at finite rates only at conditions substantially removed from equilibrium conditions (as discussed previously). Considering the hypothetical situation of figure 3.10, the reaction $X \rightarrow Y$ can run at a geologically appreciable rate only at $T_1 > T_E$, whereas $Y \rightarrow X$ can run appreciably only at $T_2 < T_E$.[7] In terms of time, $X \rightarrow Y$ can run appreciably only between t_1 and t_2, and

$Y \to X$ can run appreciably only from t_4 until the present. But, since reaction rates decrease rapidly with decreasing T (as discussed previously), and since all temperatures between t_1 and t_2 exceed all temperatures after t_4, the reaction $X \to Y$ must proceed more quickly (and hence more completely) than $Y \to X$. In fact, $Y \to X$ may not occur at all, despite the much longer time available from t_4 to the present, unless some catalytic factor, such as deformation, intervenes.

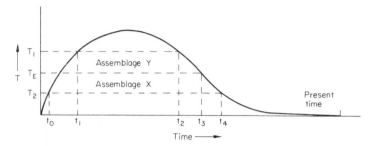

Fig. 3.10 T-time plot for a hypothetical area showing a reaction between assemblages X and Y, the equilibrium temperature being T_E. The curve represents the gradual heating and cooling of the area (assuming that only one episode of heating took place). After Fyfe, Turner and Verhoogen, p. 54.[7]

However, experiments suggest that for many metamorphic reactions (especially devaporisation reactions) $T_1 - T_E$ and $T_2 - T_E$ appear to be so small that no significant overstepping should occur.[7] If so, the foregoing explanation cannot apply, and perhaps $X \to Y$ is catalysed by factors that happened to be absent when $Y \to X$ should have occurred. Alternatively, if we are considering devaporisation reactions, once $X \to Y$ has proceeded to completion, $Y \to X$ can go only if volatile material is returned to the reaction site. This either may not be available, or, if available, might require high diffusivity paths, assisted by deformation perhaps. So the volatile material may not be able to return, in which case reaction $Y \to X$ will not proceed, regardless of how much time is available.

This leads to the general question: could two or more assemblages be forming in the same general area because catalytic factors operate in some places and not in others? Certainly this can occur in retrograde metamorphism. For example, at Broken Hill, Australia, retrograde (hydration) reactions are much more complete in local deformation zones involving extensive recrystallisation to fine-grained schistose aggregates, whereas outside these zones the same reactions are incomplete, of patchy distribution and largely pseudomorphous in the un-

deformed gneisses.[24] But is this idea applicable to prograde metamorphism? Figure 3.11 indicates that assemblage A is stable in 'zone' 1,
B in zone 2 and C in zone 3, and this progressive sequence $A \rightarrow B \rightarrow C$
should apply if equilibrium is achieved. But, once A has been formed,
C can form metastably from it between T_2 and T_3, although B is the

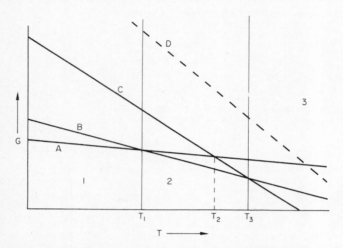

Fig. 3.11 G-T diagram, showing how mineral C could form metastably below
its lower stability temperature, and how mineral B could form metastably
above its upper stability temperature. After Fyfe, Turner and Verhoogen,
p. 55.[7]

stable assemblage, and, similarly, B could form metastably from A above
T_3, instead of C, which should form for thermodynamic reasons. As an
extreme example, all three assemblages (A, B and C) could form at any
temperature from an unstable assemblage D (of very high free energy), if
it existed in the rock. This could be the situation in experiments employ-
ing reactive (high free energy) materials, such as glass or oxide mixes.
Theoretically, it might apply to the natural prograde metamorphism of
pelitic sediments, if reactive clay minerals produced the minerals of the
various metamorphic zones by direct reactions. However, in view of the
observed relative rapidity of dehydration reactions (admittedly simplified) in the laboratory, it may be more likely that the assemblage of each
zone first goes through that of the preceding zone; i.e. that the reactions
keep pace with a slowly increasing P-T gradient.[7,22]

As mentioned in Chapter 2, diffusion limitations can restrict equilibrium to a local scale (even down to that of a few grains), and examples
of partly completed prograde and retrograde reactions in local systems

are discussed in Chapter 4. Similarly, diffusion of certain elements may be restricted to the scale of a single grain, producing chemical zoning.

Zoning in Metamorphic Minerals

Before the invention of the electron microprobe analyser, it was assumed generally that most metamorphic minerals were unzoned under most circumstances. However, increasing use of the microprobe has revealed zoning in metamorphic minerals, especially garnet. This indicates very sluggish kinetics of diffusion in many minerals under metamorphic conditions, which is suggested also by the presence of inclusions, even in minerals of high-grade rocks (Chapter 5). Zoning is usually relatively simple and preserves evidence either of changing metamorphic conditions or post-crystallisation diffusion (Chapter 4). Oscillatory zoning may occur rarely, as in grossular-rich garnet, which possibly may be explained by local oscillations in the supply of components, controlled by diffusion rates. Complex zoning in plagioclase commonly is of residual igneous origin, but rarely occurs in grains growing in cavities in metamorphic rocks (Ray Binns, personal communication), presumably owing to changes in the concentration of components in the fluid solution occupying the cavities. Metamorphic reactions involving zoned minerals are discussed in Chapter 4.

Problems of Aragonite in Metamorphic Rocks

The importance of kinetics in metamorphism is exemplified by the occurrence of aragonite (the high-pressure, low-temperature polymorph of $CaCO_3$; Fig. 4.18) in certain metamorphic rocks. In some of these rocks the problem is simply one of metastable persistence of aragonite that thermodynamically should have inverted to calcite during lowering of pressure, whereas in other rocks the problem involves metastable growth as well as metastable persistence.

Metastable Persistence Aragonite occurs in rocks of the glaucophane-lawsonite schist facies (Fig. 1.5), where it appears to be a stable mineral, on the basis of experimental and petrographic observations.[4] Thermodynamically, aragonite is unstable with respect to calcite under most crustal P-T conditions and under atmospheric conditions. We can understand the formation of aragonite in the unusually high-pressure, low-temperature gradients of the glaucophane-lawsonite schist facies, but the problem is to explain how it can survive long periods of unloading until it is exposed in these rocks at the earth's surface, especially since the aragonite → calcite transformation can be achieved readily in experiments under P-T conditions near the equilibrium curve. This must be a kinetic problem, and we need to know the reason for this metastable persistence of aragonite.

Most metamorphic aragonite shows partial inversion to calcite, the replacement involving either (i) the nucleation and growth of several calcite grains in different orientations within a single aragonite grain, or (ii) the nucleation and growth of one dendritic calcite grain within a single aragonite grain, the replacement apparently being epitaxial and involving coincidence of an *a*-axis and an [*f*:*f*] edge of calcite (both directions of closest-packing of Ca^{2+} ions) with two of the *a*, *b* or *c*-axes of aragonite.[4]

All experiments show that the rate of the aragonite → calcite transformation depends greatly on the actual aragonite sample used. For example, experimental investigation of the dry (solid-solid) reaction shows that chemical impurities have less effect than the shapes and sizes of the aragonite grains.[15] As expected, fine-grained material generally reacts more quickly than coarser-grained material, owing mainly to increased surface or interfacial area. However, geologically older aragonite transforms more slowly at a given temperature than younger aragonite, which possibly may be due to long-term, low-temperature 'annealing' that reduces the free energy of the aragonite by reducing internal strains, coalescing possible former 'domains', and increasing the degree of ordering.[15] This may assist aragonite in its persistence through long periods of geological time without inverting to calcite.

However, experiments in the presence of an aqueous fluid phase show that the solvent has a strong catalytic effect and lowers the activation energy of the transformation.[2] This is the situation in the relatively rapid conversion of aragonite to calcite in diagenetic environments, the aragonite having grown metastably (in contrast to the situation believed to apply in rocks of the glaucophane-lawsonite schist facies, which involves stable growth and metastable persistence). Whereas in the dry transformation the reaction appears to involve nucleation and growth of calcite in solid aragonite[4], the reaction in the presence of fluid involves solution and equilibration of aragonite with the solution, and the subsequent heterogeneous nucleation of calcite on the aragonite surface.[2] Various solute ions affect the reaction rate, but calcite crystallisation is most strongly inhibited by Mg^{2+} ions, which probably accounts for aragonite persistence in sea-water.[2]

So, for metamorphic aragonite to persist during slow unloading, the experimental evidence suggests that the rock must be non-porous and devoid of free water since the close of the high-pressure metamorphism that produced the aragonite. If dry, aragonite could survive unloading at temperatures of 200–300°C where gradients of only 10–12°C/km intersect the inversion curve.[4]

Metastable Growth   Metamorphic aragonite is widespread in recrystallised limestones and as veins in associated metasedimentary and metavolcanic rocks in northwest Washington, USA.[23] The problem is that other minerals present suggest metamorphism occurred under conditions of the prehnite-pumpellyite metagreywacke facies; i.e. at pressures too low for stable aragonite. Some of the aragonite is very coarse-grained and shows partial conversion to fine-grained calcite in irregular to crystallographically controlled patches or veinlets. Two interpretations are possible, namely: (i) the aragonite grew metastably, and (ii) experimentally determined equilibrium data represent metastable equilibrium. The second interpretation is always possible, as far as metamorphic reactions are concerned, but several experimenters are in reasonable agreement[22], and so we will assume that the curve of figure 4.18 is approximately right. If so, we must conclude that aragonite can grow metastably in regional metamorphism, as it can in biochemical environments (discussed above). This would be most likely to occur in deformed rocks such as the rocks in question, since experiments have shown that aragonite grows metastably in strained calcite at lower pressures than those indicated from the equilibrium phase diagram.[18]

Kinetics of the Reaction: Calcite + Quartz $\rightleftharpoons$ Wollastonite + CO$_2$

The kinetics of the reaction: calcite + quartz $\rightleftharpoons$ wollastonite + CO$_2$ have been studied in the ranges 800–950°C and *ca.* 0·07 – 1·7 kb.[14] The study involved 150 experimental runs, which gives an idea of the amount of work necessary in kinetic investigations. Pressure, temperature and grain-size of reactants were varied, to determine their effects on the reaction rate. Variations in temperature and grain-size affected the reaction rate systematically, but pressure effects were unpredictable (probably because they were masked by other kinetic factors such as variation in grain-size of the reactants).

The activation energy for the reaction was found to be between 25·5 and 29 kc/mole, which is similar to those for some other decarbonation reactions (24·5–31 kc/mole), dissociation of $CaCO_3$ to $CaO + CO_2$ (38 kc/mole), and self-diffusion of Ca in CaO (34 kc/mole). Knowing the time taken experimentally to produce a certain thickness of wollastonite in water-free conditions, Kridelbaugh[14] has calculated that rims of wollastonite (0·1–1 cm thick) that nucleated on quartz in a calcsilicate hornfels adjacent to a high-level dolerite dyke at Inverkip, Scotland, could have grown in 2·6 years at 950°C or 480 years at 800°C, under a confining pressure of 0·3 kb.

References

1 Anderson, D. E. and Buckley, G. R. (1973). Zoning in garnets—diffusion models. *Contribs. Mineralogy & Petrology*, **40**, 87–104.
2 Bischoff, J. L. and Fyfe, W. S. (1968). Catalysis, inhibition, and the calcite-aragonite problem. I. The aragonite-calcite transformation. *Amer. J. Science*, **266**, 65–79.
3 Brophy, J. H., Rose, R. M. and Wulff, J. (1964). *The Structure and Properties of Materials*. Vol. II. *Thermodynamics of Structure*, 62–111. New York: J. Wiley & Sons, Inc.
4 Brown, W. H., Fyfe, W. S. and Turner, F. J. (1962). Aragonite in California glaucophane schists, and the kinetics of the aragonite-calcite transformation. *J. Petrology*, **3**, 566–82.
5 Clemm, P. J. and Fisher, J. C. (1954). The influence of grain boundaries on the nucleation of secondary phases. *Acta Metallurgica*, **3**, 70–3.
6 Fyfe, W. S. (1964). *Geochemistry of Solids. An Introduction*, 151–69; 178–92. New York: McGraw-Hill Book Co. Inc.
7 Fyfe, W. S., Turner, F. J. and Verhoogen, J. (1958). Metamorphic reactions and metamorphic facies. *Geol. Soc. America Memoir*, **73**, 53–103.
8 Giletti, B. J. and Anderson, T. F. (1972). Diffusion of oxygen in phlogopite. *Geol. Soc. America Abstracts*, **4**, 517.
9 Green, H. W. (1972). Metastable growth of coesite in highly strained quartz. *J. Geophys. Research*, **77**, 2478–82.
10 Greenwood, H. J. (1963). The synthesis and stability of anthophyllite. *J. Petrology*, **4**, 317–51.
11 Hobbs, B. E. (1968). Recrystallization of single crystals of quartz. *Tectonophysics*, **6**, 353–401.
12 Jensen, M. L. (1965). The rational and geological aspects of solid diffusion. *Canadian Mineralogist*, **8**, 271–90.
13 Kingery, W. D. (1960). *Introduction to Ceramics*. 161–78; 217–43; 286–99; 316–40. New York: J. Wiley & Sons, Inc.
14 Kridelbaugh, S. J. (1973). The kinetics of the reaction: calcite + quartz = wollastonite + carbon dioxide at elevated temperatures and pressures. *Amer. J. Science*, **273**, 757–77.
15 Kunzler, R. H. and Goodell, H. G. (1970). The aragonite-calcite transformation: a problem in the kinetics of a solid-solid reaction. *Amer. J. Science*, **269**, 360–91.
16 Lin, T. H. and Yund, R. A. (1972). Potassium and sodium self-diffusion in alkali feldspar. *Contribs. Mineralogy & Petrology*, **34**, 177–84.
17 Moffatt, W. G., Pearsall, G. W. and Wulff, J. (1964). *The Structure and Properties of Materials*. Vol. I. *Structure*, 76–7. New York: J. Wiley & Sons, Inc.
18 Newton, R. C., Goldsmith, J. R. and Smith, J. V. (1969). Aragonite crystallization from strained calcite at reduced pressures and its bearing on ara-

gonite in low-grade metamorphism. *Contribs. Mineralogy & Petrology*, **22**, 335–48.

19 Petrovič, R. (1974). Diffusion of alkali ions in alkali feldspars. In *The Feldspars*, Ed. MacKenzie, W. S. and Zussman, J. 174–82. Manchester University Press.

20 Shewmon, P. G. (1963). *Diffusion in Solids*. New York: McGraw-Hill Book Co.

21 Sippel, R. F. (1963). Sodium self diffusion in minerals, *Geochim. et Cosmochim. Acta*, **27**, 107–20.

22 Turner, F. J. (1968). *Metamorphic Petrology. Mineralogical and Field Aspects*. New York: McGraw-Hill Book Co.

23 Vance, J. A. (1968). Metamorphic aragonite in the prehnite-pumpellyite facies, northwest Washington. *Amer. J. Science*, **266**, 299–315.

24 Vernon, R. H. (1969). The Willyama Complex, Broken Hill area. *J. Geol. Soc. Australia*, **16**, 20–55.

Chapter 4

Reactions in
Metamorphic Rocks

Introduction

Some metamorphic reactions have been mentioned as examples of
principles discussed in Chapter 2. Here I want to talk about (1) attempts
to recognise realistic reactions from microstructural and electron microprobe investigations of metamorphic rocks, and (2) attempts to relate
observation and experiment on metamorphic reactions and phase
equilibria. I will not try to cover everything, but will concentrate on a
few apparently relevant reactions in metapelitic, ultramafic and mafic
rocks, using P-T grids to give an idea of reaction sequences, where
possible, and emphasising reactions that appear to be most realistic.

The problems of recognising realistic metamorphic reactions and their
mechanisms are considered first, followed by reactions in prograde
metamorphism. Then some retrograde reactions will be considered,
followed by some reactions involving zoned minerals. Finally, local
metasomatic changes during metamorphism are considered.

Mechanisms of Metamorphic Reactions

Having considered the equilibrium aspects (Chapter 2) and the kinetic
aspects (Chapter 3) of metamorphic reactions, the next logical step is
to consider possible mechanisms by which the reactions take place. This
investigation involves two main steps: (1) to infer the actual reactions
that took place in the rock being investigated, and (2) to infer the reaction
mechanism.

(1) The first step requires the careful co-ordination of *microstructural*
interpretation (What mineral has replaced another? Where have the
various minerals nucleated? Is any microstructural evidence left or has
the reaction gone to completion?) and *chemical analysis* on the scale
of individual grains or even parts of grains, with the electron probe
micro-analyser. Unfortunately, the interpretation of microstructures

can be too subjective, and is safest where both the original mineral and its reaction products are visible, as in partial pseudomorphism, a situation much more common in retrograde than prograde situations. However, even in apparently simple reactions the mechanism may be more complex than suspected, as discussed later. In prograde metamorphism, the situation is complicated by the tendency of all phases to react and for reactions to go to completion. So, reactions can commonly be inferred only from changes in assemblages across isograds, although detailed microstructural interpretations have been used also, where evidence of partial pseudomorphism or reaction coronas is available (as discussed later for some reactions in metapelitic and mafic systems).

(2) The second step requires a time-consuming experimental investigation of the reaction, in order to work out the nucleation and growth (including diffusion) mechanisms, and has been done for very few reactions of metamorphic interest. As examples of the amount of careful experiment needed to work out the mechanism of even an apparently simple metamorphic reaction, consider the following: (*a*) the production of $MgFe_2O_4$ (a spinel) from MgO and Fe_2O_3, and (*b*) the production of wollastonite from calcite and quartz.

(*a*) *MgFe₂O₄ reaction*.[20] Consider the apparently simple reaction: $MgO + Fe_2O_3 \rightleftharpoons MgFe_2O_4$. In an experimental arrangement consisting of a block of MgO placed against a block of Fe_2O_3 (i.e. a 'diffusion-couple' arrangement), an initial nucleation of $MgFe_2O_4$ occurs at the boundary, followed by the growth of a layer of this product between the reactant blocks. Further progress of the reaction can occur only by diffusion of reactant components through the $MgFe_2O_4$ layer. The question is: which reactant ions migrate? The answer, of course, can help to explain the kinetics of the reaction. Three diffusion possibilities (i.e. reaction mechanisms) exist (Fig. 4.1; remembering that ionic diffusion implies simultaneous transport of electrical charges to maintain net electrical neutrality across the reaction layer), namely:

(i) only the cations diffuse, $3Mg^{2+}$ and $2Fe^{3+}$ migrating in opposite directions;

(ii) both cations and anions diffuse, so that $2Fe^{3+}$ and $3O^{2-}$ (or $3Mg^{2+}$ and $3O^{2-}$) migrate in the same direction; or

(iii) only the cations diffuse, but iron migrates at Fe^{2+} instead of Fe^{3+}, becoming reduced to Fe^{2+} as it leaves the Fe_2O_3 and becoming oxidised back to Fe^{3+} as it reaches the other side of the spinel layer; in this mechanism the oxygen is transported through a fluid phase, by a loss and subsequent addition of oxygen caused by oxidation-reduction reactions at the phase boundaries.

The problem is how to distinguish between these three mechanisms. It can be solved experimentally by marking the initial $MgO–Fe_2O_3$ boundary with an inert marker (e.g. a wire of some unreactive material). The marker would be displaced (owing to diffusion of material) differently for each mechanism (Fig. 4.1). It turns out that mechanism

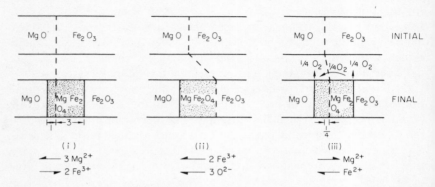

Fig. 4.1   Diffusion couple of MgO and Fe_2O_3 reacting to form a layer of $MgFe_2O_4$ by the three mechanisms shown by ion movements and discussed in the text. The initial and final position of an inert marker (dashed line) are shown for each mechanism. After Kooy, pp. 274, 275, 276.[20]

(iii) is the right one, and the explanation (based on diffusion-couple experiments and phase equilibrium diagrams) makes interesting further reading.[20]

(*b*) *Wollastonite reaction.*[12, 23] Experiments on the reaction: calcite + quartz = wollastonite + CO_2 indicate that the mechanism depends strongly on the composition of the fluid phase. In pure CO_2, the wollastonite nucleates on the quartz, whereas in a CO_2–H_2O fluid the wollastonite nucleates on the calcite. In both situations an initially rapid reaction rate is followed by a decline, indicating that nucleation of wollastonite is not a rate-determining mechanism. Instead, the rate is controlled by the rates of diffusion of reactant components through the wollastonite produced, which forms protective rims around reactant grains.

In the experiments using a CO_2–H_2O fluid, wollastonite needles nucleate at specific centres (probably defect or impurity sites) on the surfaces of calcite grains and then grow outwards as rosettes, mainly parallel to the calcite surface, thereby rapidly forming a reaction rim.[12] In the experiments using pure CO_2, the wollastonite is granular, not acicular, and forms rims on quartz particles.[23] However, this does not appear to have involved a simple volume replacement of quartz by wollastonite, because of an observed Si concentration gradient between the rim of wollastonite and the interior of the quartz grain. A Ca gradient also occurs across the reaction rim, so that diffusion of both Si and Ca appears to have taken place, presumably accompanied by diffusion of oxygen to maintain electrical neutrality. Probably the diffusion of Ca and Si in opposite directions causes the wollastonite to grow both on the interior and exterior of the reaction rim. Electron microprobe scans (Fig. 4.2) across initial calcite-quartz contacts (now separated by

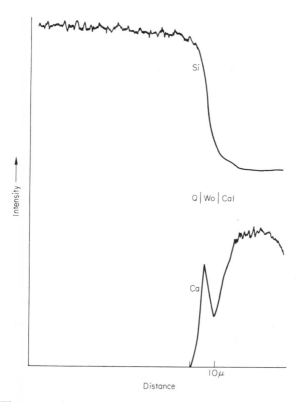

Fig. 4.2   Electron microprobe scans for silicon and calcium across quartz–wollastonite and wollastonite–calcite grain boundaries. Note the rapid drop in calcium at the wollastonite–calcite boundary and the increase in the wollastonite rim. After Kridelbaugh, p. 766.[23] Cal=calcite; Q=quartz; Wo= wollastonite.

wollastonite reaction rims) show a concentration gradient between the wollastonite and calcite, suggesting that diffusion of Ca and oxygen is maintained by a concentration gradient in these local dry systems.[23] The effect of the composition of the fluid phase on the reaction mechanism is connected with the differing solubilities of calcite and quartz under differing conditions.[23] Calcite is more soluble with increasing X_{CO_2} (allowing Ca to diffuse in ionic form through the fluid to the reaction interface, so that wollastonite nucleates on the relatively insoluble quartz), whereas quartz is more soluble with increasing X_{H_2O}.

On the basis of the above experimental observations, and assuming a fluid phase to be present, calcsilicate rocks showing evidence of wollastonite nucleating on and replacing clastic quartz grains should imply high X_{CO_2} in the fluid,[23] whereas wollastonite replacing calcite fossils should imply a significantly lower X_{CO_2} (higher X_{H_2O}) in the fluid.

The situation in an actual calcsilicate rock may be somewhat different from that in the experiments, although enough porosity might be expected for a fluid phase to exist, at least in the earlier stages of reaction (Chapter 2). Many calcite/quartz contacts should occur, however. Presumably the reaction mechanism would partly involve solution of one phase and deposition of wollastonite on the other, followed by advancement of the wollastonite interface both into the phase acting as a nucleating substrate and into the fluid-filled spaces. Also, we might expect it to advance into adjacent solid grains of the other reactant phase by the dry diffusion mechanism mentioned above.

The effect of reduced grain-size on increased reaction rates could be due to (*a*) increased surface area for solution of the more soluble reactant phase, coupled with (*b*) increased number of nucleation sites and area for growth of the product phase on grains of the less soluble reactant phase.[12]

The observed slowing down of reactions, owing to the formation of a layer of products between reactant grains, should be considered as a possibility in natural metamorphic rocks, and may explain some occurrences of wollastonite, quartz and calcite in calcsilicate rocks. Of course, other explanations are possible also (Chapters 2, 3).

Reactions in Metapelitic Systems

I will now discuss some reactions in systems of aluminous composition, beginning with the simplest system and progressing into increased complexity.

System: Al_2SiO_5   This famous system has been studied by many people, with embarrassingly different results (Fig. 4.3). As discussed previously, reactions in this system are driven by only small free energy changes, and involve large activation energies (caused by the necessity to break strong Si-O and Al-O bonds). Therefore, overstepping and metastability are problems in experimental location of the reaction curves, and probably are common in rocks also. For example, natural occurrences of all three polymorphs are fairly common, which implies that (i) metamorphic P-T conditions coincided with the triple point (which is unlikely, though possible); (ii) one or two of the polymorphs was metastable; or

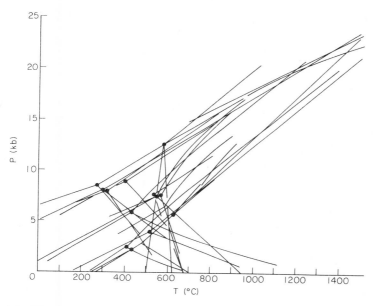

Fig. 4.3   The Al_2SiO_5 reaction curves in positions inferred by various experimenters. After Althaus (1969) *Neues Jahrb. für Mineralogie*, **111**, p. 114.

(iii) the reaction curves are not univariant, as commonly supposed, but divariant, owing to substitution of small amounts of, say, Fe^{3+} for Al^{3+}; for example, Strens has suggested that kyanite-sillimanite and andalusite-sillimanite can coexist stably over pressure ranges of 0·1 to 0·4 kb[30]; but Holdaway concluded that Fe^{3+} has a negligible effect, except for andalusite $\rightleftharpoons$ sillimanite (to a small extent).[17] Another complicating factor is the suggestion of Greenwood and others that disorder of tetrahedral Si and Al in sillimanite, even in amounts too small to measure by common X-ray techniques, may affect the Al_2SiO_5 phase relationships almost as much as solid solution of minor elements.[14]

We need to know the position of the Al_2SiO_5 triple point, because it limits the upper pressure of natural andalusite-bearing assemblages (i.e. of low-pressure metamorphism; Fig. 1.1). Happily, some of the experimental data of figure 4.3 can be reconciled with thermodynamic calculations, and a reasonable phase diagram has been produced (Fig. 4.4), placing the triple point at about 490–540°C and 3·7–4·3 kb. The experimental determinations used in figure 4.4 are those obtained with hydrothermal or gas-pressure equipment. The latter was not described previously (Chapter 2), because it is relatively uncommon, owing to

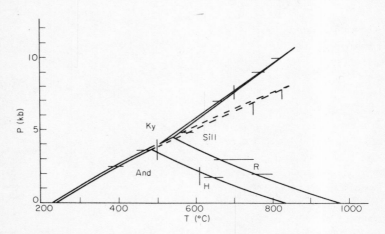

Fig. 4.4   Phase equilibrium diagram of the Al_2SiO_5 system, showing experimental reversals (long dashes) obtained with hydrostatic pressure media and relatively pure starting materials. Also shown are curves calculated by Wall and Essene (1972, *Geol. Soc. America Abstracts*, **4**, p. 700), neglecting any possible effects of Al-Si order/disorder. The dashed lines are metastable extensions of the andalusite ⇌ kyanite reaction into the stable sillimanite field. The calculations, within their order of accuracy, fit either of the two sets of andalusite ⇌ sillimanite experimental reversals shown, namely R (Richardson, Gilbert, and Bell, 1969, *Amer. J. Science*, **261**, pp. 259–72) and H (Holdaway[17]).

technological difficulties involved with its construction and maintenance. However, it applies a true and accurately measurable hydrostatic pressure to the sample, which removes the kinetic (and possibly thermodynamic) effect of differential stress and the kinetic effect of plastic deformation of reactants during the experiment. This effect can be a major factor in the metastable crystallisation of phases in polymorphic transformations, especially where ΔGs are small, so that the free energy introduced by deformation may be of a similar order to that of the polymorphic change itself.[3]

Pure Al_2SiO_5-rocks are rare, and probably are due mainly to the metamorphism of leached residual deposits. Water, alkalis and other components are present in most metapelitic rocks, so that the simple

Al$_2$SiO$_5$ polymorphic changes generally do not occur. For instance, kyanite is stable at low temperatures (Fig. 4.4), but does not occur in metapelites of the greenschist facies, its place being taken by hydrous phases, as explained below. Moreover, the apparently simple polymorphic change from kyanite to sillimanite in metapelitic rocks probably is complicated, commonly involving other minerals (Chapter 2).

System: Al$_2$O$_3$-SiO$_2$-H$_2$O Adding water to the Al$_2$SiO$_5$ system converts the P-T phase diagram to that shown in figure 4.5. The following points are noteworthy:

(i) The Al$_2$SiO$_5$ and quartz ⇌ coesite polymorphic equilibria are shown. Note that quartz ⇌ coesite is strongly pressure-dependent, and that coesite is characteristic of impact metamorphic environments. However, it can grow in strongly plastically deformed quartz up to 12 kb below its thermodynamic hydrostatic inversion pressure at 900°C,[16] which is a striking example of the kinetic effect of plastic deformation in polymorphic transformations.

(ii) The presence of water prevents the formation of kyanite at low temperatures, minerals such as pyrophyllite and diaspore appearing instead.

(iii) Both the lower- and upper-temperature stability limits of pyrophyllite are shown. Note that these lie outside the P-T limits of the pyrophyllite-diaspore assemblage, as expected from previous discussions (Chapter 2). Similarly, kyanite is stable to lower temperatures than kyanite + quartz, and kaolinite is stable to higher temperatures than kaolinite + quartz.

(iv) The assemblage diaspore-quartz should be characteristic of the glaucophane-lawsonite schist facies (Fig. 1.5), although it would be expected to be rare, owing to rarity of rocks of suitable composition (e.g. metamorphosed bauxites). In fact, it has been observed in some glaucophane-lawsonite terrains.

(v) Pyrophyllite is stable over a maximum temperature range of only about 100°C, so that the temperature range of assemblages such as quartz-chlorite-pyrophyllite (which might be expected to occur in some metapelitic rocks), would be still smaller.[31] This, as well as the relative rarity of alkali-poor pelitic compositions, severely restricts the occurrence of pyrophyllite in rocks. However, where present, it is a good indicator of the greenschist or albite-epidote hornfels facies.

(vi) The two reaction curves involving kaolinite dehydration show notable change of slope ('back-bending') as discussed previously for analcite dehydration (Chapter 2).

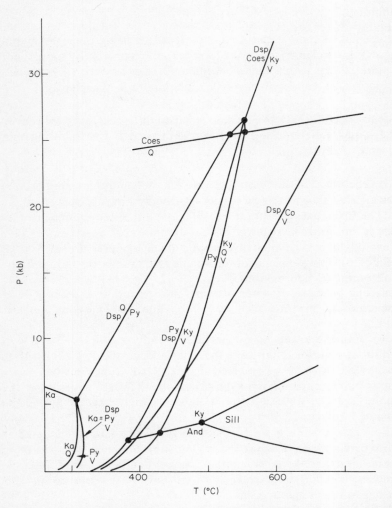

Fig. 4.5   Phase equilibria in the system Al_2O_3–SiO_2–H_2O calculated by Wall and Essene (1972, *Geol. Soc. America Abstracts*, **4**, p. 700). And=andalusite; Co=corundum; Coes=coesite; Dsp=diaspore; Ka=kaolinite; Ky=kyanite; Py=pyrophyllite; Q=quartz; Sill=sillimanite; V=water vapour.

System: Al_2O_3-SiO_2-H_2O-Na_2O-K_2O   Reactions involving the breakdown of paragonite and quartz+paragonite are analogous to those for breakdown of muscovite and quartz+muscovite (Fig. 2.5), but take place at lower temperatures (less than 650°C at pressures below 7 kb).[6,7] Paragonite-quartz assemblages are relatively common in certain aluminous medium-grade rocks (occurring with minerals such as staurolite, kyanite, andalusite and garnet), but break down well before muscovite-quartz in progressive metamorphism. However, paragonite can be stabilised to higher temperatures by solid solution of the muscovite component. Generally, Na is low enough and K high enough in metapelitic rocks for a white mica solid solution to form instead of paragonite, although in some relatively K-poor rocks both muscovite and paragonite coexist. In normal pelitic compositions, paragonite probably disappears by (i) complex reactions producing garnet, staurolite and kyanite, and (ii) progressive substitution of Na into muscovite.[9] This increased mutual solubility of paragonite and muscovite at higher temperatures could be a useful geothermometer with more experimental data, as it can be correlated with the d_{002} spacing of the mica.[9]

The upper temperature limit of muscovite+quartz occurs within the upper hornblende hornfels and amphibolite facies (Fig. 1.5), and is generally correlated with the appearance of sillimanite+K-feldspar (Fig. 2.5). In western Maine, USA, Evans and Guidotti[9] inferred the following reaction (on the basis of observed assemblages and electron microprobe analyses of minerals) marking the sillimanite-orthoclase isograd:

$$K_{0.94}Na_{0.06}Al_2AlSi_3O_{10}(OH)_2 + SiO_2 + 0.1(NaAlSi_3O_8) \rightleftharpoons$$
muscovite   quartz   albite component
                                  of plagioclase

$$1.1(K_{0.86}Na_{0.14}AlSi_3O_8) + Al_2SiO_5 + H_2O$$
orthoclase   sillimanite

This is a more realistic reaction for many rocks than the simplified reaction without sodium (Fig. 2.5). The reaction is discontinuous, because $C=5$, $P=6$; hence $F=1$. Therefore, ideally it would not be expected to occur over a significant grade interval. Yet, in western Maine the assemblage, quartz + muscovite + orthoclase + sillimanite + plagioclase, persists for at least seven miles from the sillimanite-orthoclase isograd. The fact that this is not due to reaction-sliding is supported by the uniformity of orthoclase and muscovite compositions over the seven miles. Possible explanations are as follows:

(i) Although most dehydration reactions are said to be relatively rapid, especially in rocks, the quartz-muscovite dehydration reaction is known to be slow in experiments, owing not to nucleation, but to the slow growth rate of sillimanite.[8] So, this kinetic explanation is possible, although not very likely, because, since the reaction is well advanced a mile from the isograd, the question arises as to why it isn't complete a further six miles away.

(ii) If the rocks were sufficiently impermeable to restrict the escape of H_2O produced, the reaction would slow down (Chapter 2), its rate being controlled by the rate at which H_2O was removed from the reaction zone.[9] This could account for both the widespread coexistence of reactants and products, and the uniformity of mineral composition.

System: Al_2O_3-SiO_2-H_2O-Na_2O-K_2O-MgO-FeO-Fe_2O_3   This is virtually the complete 'pelitic system', and so is of the greatest relevance to actual rocks. Unfortunately, the multiplicity of components makes experiments difficult to carry out, so that we have little experimental information in this system on realistic reactions. Some information is available on realistic minerals, but many of the reactions studied are simplified.

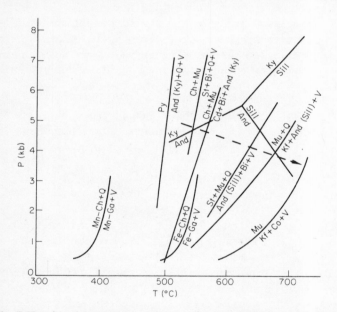

Fig. 4.6   P-T grid of several experimentally determined equilibria relevant to metapelitic rocks, with P-T gradient (dashed line) inferred for the aureoles of the Donegal granites. After Naggar and Atherton (1970). *J. Petrology*, **11**, p. 582, © 1970 Oxford University Press, by permission of The Clarendon Press, Oxford. The curves shown were valid at the time the grid was constructed, and no attempt has been made to bring them up to date, because only a general point is being made (see text). And=andalusite; Bi=biotite; Cd=cordierite; Ch=chlorite; Co=corundum; Ga=garnet; Kf=K-feldspar; Ky=kyanite; Mu=muscovite; Py=pyrophyllite; Q=quartz; Sill=sillimanite; St=staurolite; V=water vapour.

However, 'petrogenetic grids' can be drawn up from available data, allowing reasonable estimates of P-T gradients in some areas (Figs 4.6, 1.2).

Of course, another problem is that, because of the large number of phases and components, reactions are difficult to infer from the rocks themselves, and, since this kind of information is required in order to design realistic experiments, the problem is made doubly difficult. As pointed out by Turner,[31] the formation of the common low-grade assemblage: quartz-white mica (commonly phengitic)-chlorite (commonly aluminous)-albite, from unstable detritus (mainly clay minerals, quartz, and feldspar), is poorly understood. The detailed nature of the reactions, and consequently their P-T-X conditions, are difficult to infer or study experimentally, especially since the reactants themselves probably do not constitute a stable assemblage.

At higher grades, chlorite and white mica react to form biotite, and detailed field, mineralogical and chemical work has enabled the inference of several apparently realistic reactions. Examples involving stilpnomelane have been discussed already (Chapter 2). The reaction inferred by Mather[25] for metapelitic rocks in the Scottish Highlands is a particularly good example of the inference of realistic reactions from detailed mineralogical-chemical data. Before his investigation, several simple reactions had been suggested by various people for metapelitic rocks, namely:

$$\text{K-rich muscovite} + \text{chlorite} + \text{quartz} \rightleftharpoons \text{biotite} + \text{less K-rich muscovite} + H_2O$$

$$3 \text{ muscovite} + 5 \text{ prochlorite} \rightleftharpoons 3 \text{ biotite} + 4 \text{ Al-rich chlorite} + 7 \text{ quartz} + 4 H_2O$$

$$\text{phengite} + \text{chlorite} \rightleftharpoons \text{muscovite} + \text{biotite} + SiO_2 + H_2O$$

However, Mather found that the actual minerals in the Scottish Dalradian rocks do not agree with these reactions. He found also that biotite is absent from many metapelitic rocks in the classic 'biotite zone'. He investigated the associated metagreywackes and inferred the following reaction (shown graphically in Fig. 4.7):

$$\text{microcline} + \text{chlorite} + \text{phengitic muscovite} \rightleftharpoons \text{biotite} + \text{less phengitic muscovite} + \text{quartz} + H_2O$$

Now, chemical analysis of the minerals involved shows that the extent of phengitic substitution in muscovite decreases with increasing grade (Fig. 4.8). This means that, at low grades, biotite can form only in Al-poor rocks (namely metagreywackes), because in more pelitic compositions (i.e. above AX in Fig. 4.8) the pair: phengite-chlorite is still stable. As the grade increases, more Al-rich rocks can react, but in very aluminous rocks (i.e. above AB in Fig. 4.8) the appearance of biotite may be delayed

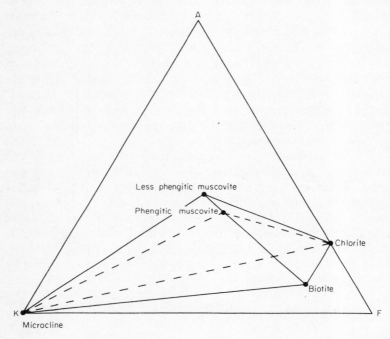

Fig. 4.7 Graphical representation (diagrammatic only) of the breakdown of the assemblage microcline-chlorite-phengitic muscovite (stable in the Dalradian chlorite zone, as indicated by dashed tie-lines) to form the assemblage biotite-less phengitic muscovite (stable in the the biotite zone, as indicated by solid tie-lines). The biotite-bearing assemblage will also contain chlorite if the bulk rock composition lies in the chlorite-biotite-muscovite triangle, or microcline if the bulk rock composition lies in the microcline-biotite-muscovite triangle. After Mather (1970). *J. Petrology*, **11**, p. 265, © 1970 Oxford University Press, by permission of The Clarendon Press, Oxford.

until even the garnet isograd has been reached. So these rocks do not carry biotite, even though they are in the biotite zone, which is a striking example of the influence of bulk chemical composition on the mineral assemblage.

An important consequence of Mather's work is that a restricted rock composition may be preferable to others for the delineation of isograds. If metagreywackes are used, an unequivocal biotite isograd based on the discontinuous breakdown of a chlorite-zone assemblage can be located. However, if metapelitic rocks are used, a particular pelitic composition must be selected, in which the appearance of biotite represents an arbitrary stage in the expansion of the phengite-chlorite-biotite field.[25]

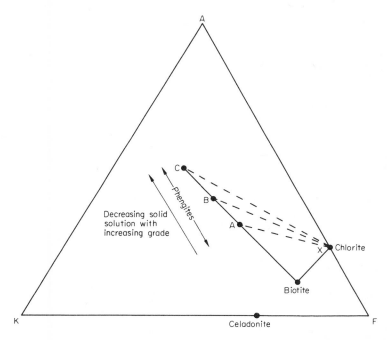

Fig. 4.8 Diagrammatic illustration of the expansion of the phengite+biotite +chlorite field with increasing metamorphic grade in the Scottish Dalradian. The limits of solid solution in the upper chlorite, biotite and garnet zones are represented by AX, BX and CX, respectively. After Mather (1970). *J. Petrology*, **11**, p. 266, © 1970 Oxford University Press, by permission of The Clarendon Press, Oxford.

Higher-grade reactions in the progressive metamorphism of pelitic rocks involve garnet, micas, aluminosilicates, staurolite, feldspars and cordierite. Many reactions have been suggested, but, once again, complexity makes recognition of real reactions and correlation between experiment and observation difficult. Some examples of detailed attempts to infer realistic reactions at these grades in metapelitic rocks follow:

(i) Carmichael has made some ingenious contributions towards a more realistic understanding of metamorphic reactions, emphasising that even apparently simple reactions may have complex mechanisms.[5] He reasoned that metamorphic reactions may involve successive steps (as is common in many experimentally investigated chemical reactions) that work in internally consistent *cycles* ('cyclic reactions'), and may leave little or no evidence of their operation.

Carmichael used as an example the apparently simple reaction kyanite $\rightleftharpoons$ sillimanite, which traditionally marks the sillimanite isograd in the Scottish Dalradian. If this were a simple pseudomorphous reaction in the Dalradian rocks, the sillimanite should occur largely in or around former kyanite grains. However, most of the sillimanite occurs in biotite and quartz. So, Carmichael looked for other possible reactions, based partly on microstructural inferences, such as:

$$3 \text{ kyanite} + 3 \text{ quartz} + 2K^+ + 3H_2O \rightleftharpoons 2 \text{ muscovite} + 2H^+ \qquad (1)$$

This reaction uses Al and Si from solid phases and K^+ and H_2O from a 'dispersed phase'. The muscovite produced by reaction (1) or other possible reactions[5] could react in the following way:

$$2 \text{ muscovite} + 2H^+ \rightleftharpoons 3 \text{ sillimanite} + 3 \text{ quartz} + 2K^+ + 3H_2O \qquad (2)$$

If reactions (1) and (2) proceeded together in the same small volume of rock, the net reaction would be the simple conversion of kyanite to sillimanite:

$$(1) + (2) = 3 \text{ kyanite} \rightleftharpoons 3 \text{ sillimanite}$$

The situation is shown diagrammatically in figure 4.9(*a*), which indicates the actual number of ions that would have to diffuse (probably along grain boundaries) between the sites of reactions (1) and (2), which can be regarded as local systems, for three moles of kyanite to become three moles of sillimanite. Each reaction acts both as a source and a sink of mobile components required and produced, respectively, by the other reaction. Therefore, no external reservoirs of material are needed, if both reactions go at the same rate, and we are dealing with a closed local system, which could be on the scale of several grains. Note, especially, that sillimanite can grow at the expense of kyanite without being in contact with it and without transport of Al or Si (so that silicate frameworks can remain intact); only soluble ions and water need move.

Carmichael then showed how the same reaction might be even more complex.[5] Although prevailing opinion is that sillimanite post-dates the biotite in which it commonly occurs in the Dalradian rocks, such a reaction involving no other phases would require introduction of

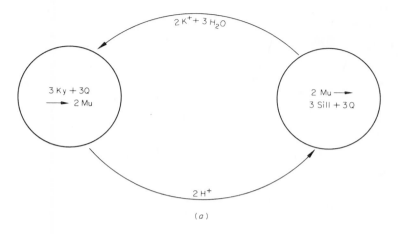

(a)

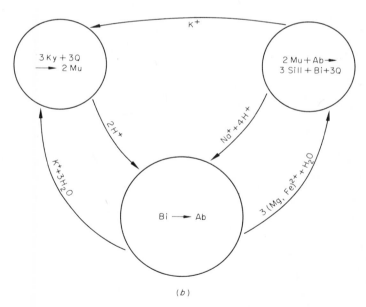

(b)

Fig. 4.9   Diagrammatic representation of two possible cyclic reaction schemes for the net reaction: Ky⇌Sill. After Carmichael (1969). *Contribs. Mineralogy and Petrology*, **20**, pp. 253, 255. Ab=albite; Bi=biotite; Ky=kyanite; Mu= muscovite; Q=quartz; Sill=sillimanite.

aluminium into the rock from an external source. Carmichael suggested the biotite and sillimanite may have formed simultaneously, which interpretation is equally valid on the basis of commonly ambiguous microstructural evidence. He postulated the following reactions:

$$\text{biotite} + Na^+ + 6H^+ \rightleftharpoons \text{albite} + K^+ + 3(Mg,Fe)^{2+} + 4H_2O \tag{3}$$

and

$$2\ \text{muscovite} + \text{albite} + 3(Mg,Fe)^{2+} + H_2O \rightleftharpoons$$
$$\text{biotite} + 3\ \text{sillimanite} + 3\ \text{quartz} + K^+ + Na^+ + 4H^+ \tag{4}$$

Reaction (3) is suggested by the (not overconvincing) evidence of plagioclase apparently corroding biotite, and the plagioclase produced could react with the muscovite produced in reaction (1) according to reaction (4). Combining reactions (1), (3) and (4), we have:

$$(1) + (3) + (4) = 3\ \text{kyanite} \rightleftharpoons 3\ \text{sillimanite},$$

which is the same net reaction as before. Once again, no change in the bulk composition of the rock is involved and, above all, no aluminium need be introduced into the rock. In fact, the only materials that need move are the ionic H^+, K^+, Na^+ and $(Mg,Fe)^{2+}$, along with water, and the system of sources and sinks is internally consistent with a local closed system (Fig. 4.9b). The Al-Si-O frameworks of the participating minerals need not be disturbed.

These cyclic reactions involve local metasomatism but no bulk chemical change, and, furthermore, the metasomatic changes are separated only in space, not in time. Most of the minerals in the rock could be involved, the overall process being driven by the negative ΔG of the net reaction. The foregoing discussion suggests that even apparently simple reactions can be very complicated and that most or all phases in a rock may participate, especially in prograde metamorphism.

(ii) As a result of a detailed electron microprobe and microstructural study of the apparent replacement of staurolite by 'coronas' of various minerals, Kwak inferred a number of prograde staurolite-breakdown reactions.[24] Some of the microstructural criteria (which are notoriously difficult to interpret in prograde metapelitic rocks) seem debatable, but one reaction that appears to be justified by the microstructural evidence is:

staurolite + garnet $\rightleftharpoons$ sillimanite + biotite + plagioclase.

Detailed electron microprobe analysis of inferred reactants and products from one particular partial pseudomorph suggests that the actual reaction at that site is:

2·14 staurolite + 2·13 garnet + 18·80 Si + 0·47 Ti + 3·12 Mg + 4·85 Ca + 5·61 K + 9·87 Na (+46·07 oxygen + 7·60 hydroxyl) $\rightleftharpoons$ 0·62 sillimanite + 3·10 biotite + 3·77 plagioclase + 10·16 Al + 6·97 Fe + 1·92 Mn + 0·40 Zn.

The various partly replaced staurolite relics through the rock are chemically relative homogeneous, whereas different grains of biotite and plagioclase through the rock have different chemical compositions. The edges of the garnet grains also are variable in composition. All these features suggest that equilibrium was not achieved on the 'rock' scale (e.g. on the scale of a normal sized thin section), but only on the scale of a single staurolite relic and its immediate replacement corona of product minerals, implying very local volumes of equilibrium (Chapter 2).

The above reaction shows (as do other staurolite reactions inferred by Kwak[24]) that Si must be available as a reactant (possibly from the free quartz present in all the rocks concerned) and that Al and Zn are produced in excess of the amounts taken up by crystalline products. A mineral 'sink' for Al at the reaction site is not obvious, implying that Al is a mobile component on the scale of the local system, regardless of whether or not an Al_2SiO_5 mineral is available to accept it in the surrounding rock. The same applies to Zn, except that no minerals in these rocks have compositions indicating that they could have accepted all the Zn produced in the reactions (staurolite being relatively Zn-rich), which suggests that it must have migrated right out of the 'rock' (i.e. it was mobile through relatively large rock volumes).

Other important points about these reactions are that different proportions of product minerals occur in different partial pseudomorphs through the rock, and that different extents of reaction occur, some staurolite grains remaining intact. These differences do not satisfactorily account for the chemical imbalances in the above specific reaction, so that the imbalances apparently reflect the existence of local systems open to various excess elements in the reaction. The third dimension has not been considered, of course, but its effect probably would not alter these inferences for the rock concerned.[24]

(iii) Finally, let us look at a complex garnet-forming reaction. In a detailed electron microprobe study of metapelitic rocks from a structurally well-known locality in British Columbia, Jones inferred the following reaction for the production of garnet:[19]

5·31 ferroan dolomite + 8·75 paragonite + 4·80 pyrrhotite + 3·57 albite (An_1) + 16·83 quartz + 1·97 O_2 ⇌ 1·00 garnet + 16·44 andesine (An_{32}) + 1·53 chlorite + 2·40 S_2 + 1·90 H_2O + 10·62 CO_2.

The coefficients of this reaction are very sensitive to the Mn content of the ferroan dolomite, which is quite variable. If it is lower than average, more ferroan dolomite would be needed to produce enough Mn for garnet, which would increase the amount of andesine produced. A schematic representation of the reaction is shown in figure 4.10, which gives a fair idea of how complicated some metamorphic reactions may be!

Reactions in Mafic and Ultramafic Systems

In view of the fact that many metamorphic facies are distinguished on the basis of mineral assemblages in rocks of mafic composition, it is unfortunate that more experimental work is not available on the relevant reactions. However, some experimental investigations merit discussion.

System: MgO-SiO_2-H_2O The main experimentally determined and calculated reaction curves in this system are shown in figures 4.11 and 4.12, respectively. Note that the upper stability temperature of serpentine is around 500°C, but in the presence of brucite it decomposes at lower temperatures. The occurrence in ultramafic rocks of anthophyllite and talc is restricted by their relatively SiO_2-rich compositions. In fact, the stabilities of serpentine, talc and anthophyllite depend on H_2O and SiO_2 activities (Fig. 4.13), as well as pressure and temperature. This explains why talc in alpine serpentinites is commonly restricted to alteration zones around quartz-bearing tectonic inclusions. Figure 4.12 shows that anthophyllite gives way to enstatite and talc above a certain pressure (6–8 kb here, although some calculations place the reaction at above 20 kb).

Addition of CO_2 to this system causes the appearance of magnesite at temperatures below 600°C at $P_{fluid} = 2$ kb (Fig. 4.14). The assemblage serpentine-magnesite is bounded by reactions A, C and F, and is severely restricted to low X_{CO_2} environments. If the CO_2 content of the fluid phase exceeds 2 to 6 mol. per cent, serpentine reacts with CO_2 to give talc + magnesite (or quartz + magnesite at lower temperatures). These reactions are relevant to serpentinites containing secondary magnesite.

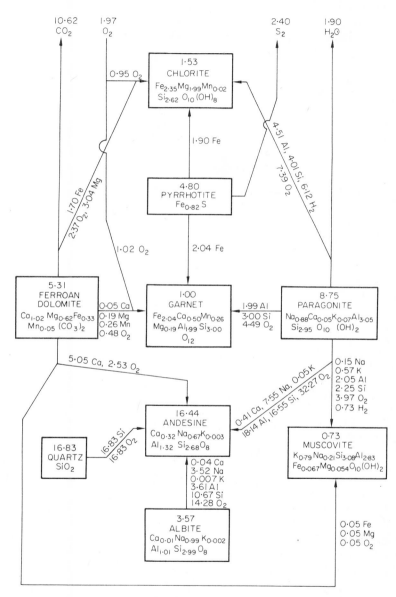

Fig. 4.10 Schematic representation of the complex garnet-forming reaction proposed by Jones, p. 298.[19]

Addition of CaO to the system results in the appearance of tremolite and diopside (Fig. 4.12). Tremolite has an upper pressure limit of 9–20 kb, and an upper temperature limit of about 750–930°C, but diopside is stable to higher pressures and temperatures.

System: MgO-SiO_2-H_2O-CaO-Al_2O_3-FeO-Fe_2O_3 This system is virtually the full 'mafic' system, and, as with the full 'pelitic' system, correla-

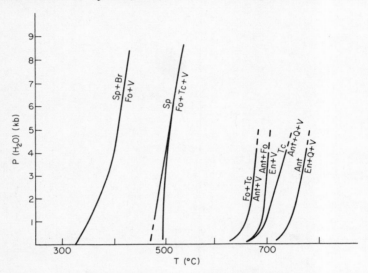

Fig. 4.11 Experimentally determined equilibrium curves for reactions in the system: MgO-SiO_2-H_2O, as summarised by Johannes (1968). *Contribs. Mineralogy and Petrology*, **19**, ṗ. 314. Ant=anthophyllite; Br=brucite; En= enstatite; Fo=forsterite; Q=quartz; Sp=serpentine; Tc=talc; V=water vapour.

tion between observation and experiment, as well as recognition of realistic reactions, is difficult. Reactions at low grades may involve hydration of initially mainly anhydrous igneous minerals (e.g. plagioclase, pyroxene, olivine) but with progressive metamorphism the reactions essentially involve dehydration of hydrated assemblages formed at lower grades. Many reactions have been suggested for mafic assemblages, but few are adequately supported by chemical analysis of the actual minerals or experimental determination of the inferred reactions.

Some of the potentially most reliable reactions in mafic rocks can be inferred from relatively fine-grained coronas (reaction rims) formed by partial reactions between coarse-grained minerals. These reactions are commonly anhydrous, which may explain why they have not gone to completion (owing to relatively slow diffusion rates through the layer

of product minerals in the absence of hydroxyl).[26] Detailed microstructural and electron microprobe investigations have revealed a large

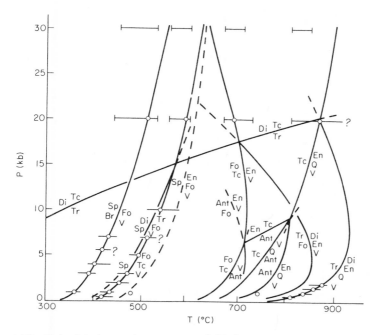

Fig. 4.12 Calculated reaction curves, of likely relevance to natural rocks,
in the system $MgO\text{-}CaO\text{-}SiO_2\text{-}H_2O$. Calculations by Essene, Wall and Shettel
(1973). *Trans. Amer. Geophys. Union*, **54**, p. 480. Ant=anthophyllite; Br=
brucite; Di=diopside; En=enstatite; Fo=forsterite; Q=quartz; Sp=serpen-
tine; Tc=talc; Tr=tremolite; V=water vapour.

variety of layered corona structures, and various reactions and reaction
sequences have been proposed to explain them. Some examples are:[15]

olivine+plagioclase ⇌ orthopyroxene+clinopyroxene+spinel
orthopyroxene+clinopyroxene₁+spinel+plagioclase ⇌
                                     clinopyroxene₂+garnet
orthopyroxene+plagioclase (±spinel) ⇌ garnet±clinopyroxene
                                        ±quartz
igneous clinopyroxene+plagioclase ⇌ metamorphic clinopyroxene
                                        +garnet
omphacite ⇌ diopside+plagioclase
garnet+olivine ⇌ orthopyroxene+clinopyroxene+spinel

The last reaction explains the occurrence of fine symplectic rims around garnet grains in many Norwegian garnet peridotites, and has been interpreted as resulting from a change from eclogite to granulite facies conditions after emplacement of the peridotite bodies into crustal

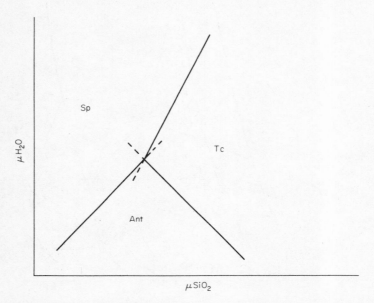

Fig. 4.13   Calculated relative stabilities of serpentine, talc and anthophyllite as functions of the chemical potentials of SiO_2 and H_2O at constant pressure and temperature. After Greenwood (1963). *J. Petrology*, **4**, p. 348, © 1963 Oxford University Press, by permission of The Clarendon Press, Oxford.

rocks.[4, 15] The others appear to be incomplete reactions in dry olivine + plagioclase and pyroxene + plagioclase assemblages during cooling from igneous temperatures in the continental crust.[15] In the sense that they result from decreasing temperature (as suggested by geological and experimental evidence), the reactions technically are 'retrograde', although assemblages of the granulite and eclogite facies (Fig. 1.5) are produced.[15]

Variation in the coronas may depend a lot on the prevailing P-T conditions.[15] Cooling of the original igneous rock at high crustal levels (i.e. at low P and T) will not permit rapid enough nucleation or growth to cause reaction, so that no coronas will be produced. Cooling at greater depth may produce fibrous (spherulitic) coronas, probably because of relatively slow diffusion (at the relatively low temperatures concerned) of components unwanted by the growing grains, a condition known to favour spherulitic growth in polymeric organic materials. Cooling at greater depths tends to produce less fibrous, more granular coronas, probably owing to more rapid removal of unwanted impurities, slower nucleation rates and the increased possibility of later recrystallisation of initially fibrous grains.[15]

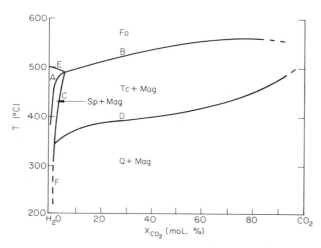

Fig 4.14 Experimentally determined T-X phase equilibrium diagram for the system $MgO-SiO_2-H_2O-CO_2$ at $P_{fluid}=2$ kb. The reactions are:

(A) $2Fo+2H_2O+CO_2 \rightleftharpoons Sp+Mag$
(B) $4Fo+H_2O+5CO_2 \rightleftharpoons Tc+5Mag$
(C) $2Sp+3CO_2 \rightleftharpoons Tc+3Mag+3H_2O$
(D) $Tc+3CO_2 \rightleftharpoons 4Q+3Mag+H_2O$
(E) $Sp \rightleftharpoons Fo+Tc+H_2O$
(F) $Sp+3CO_2 \rightleftharpoons 2Q+3Mag+2H_2O$

After Johannes (1967). *Contribs. Mineralogy and Petrology*, **15**, p. 244. Fo= forsterite; Mag=magnesite; Q=quartz; Sp=serpentine; Tc=talc.

As in some of the inferred staurolite reaction rims discussed previously, the reactions in mafic coronas may involve chemical imbalances (Fig. 4.15), so that the local system appears to be open to the excess components.[26]

Some of the most detailed experimental work on mafic and related systems concerns reactions relevant to the basalt $\rightleftharpoons$ eclogite transformation (Figs 4.16, 4.17) and the origin of 'blueschist' (glaucophane-lawsonite schist facies) metamorphism (Fig. 4.18). Extending into the area of low grade regional rock alteration ('burial' and 'hydrothermal' metamorphism), recent experimental work has enabled the construction of a P-T grid (Fig. 4.19), involving selected reactions in various mafic, impure calcareous and sodic compositions (mainly metatuffaceous rocks).

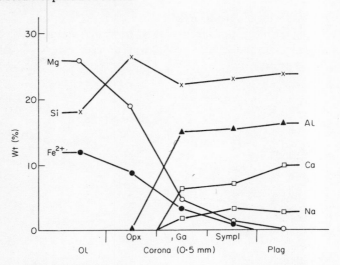

Fig. 4.15   Variation in reacting elements across a compound layered reaction corona between olivine (left) and plagioclase (right). After Miller, p. 337.[26] 'Sympl' refers to a fine-grained symplectic intergrowth. Ga=garnet; Ol= olivine; Opx=orthopyroxene; Plag=plagioclase.

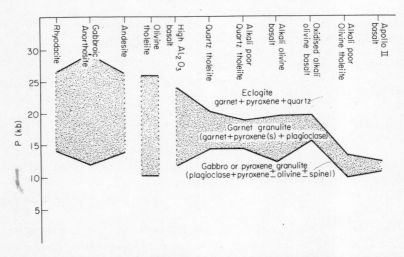

Fig. 4.16   Diagram showing the effect of changing bulk chemical composition on the pressure required for the incoming of garnet (lower boundary) and out-going of plagioclase (upper boundary) in the basalt ⇌ eclogite transformation; based on experimental data at 1 100°C. Eclogite mineral assemblages should be stable in dry basaltic rocks under normal geothermal gradients in the stable continental crust. After Green and Ringwood (1972). *J. Geology*, **80**, p. 280, © The University of Chicago. All rights reserved.

How Relevant Are Available Experimental and Theoretical Determinations of P-T-X Equilibria?

They are only as relevant as the reactions with which they are concerned. We can invent all combinations of phases and components, and con-

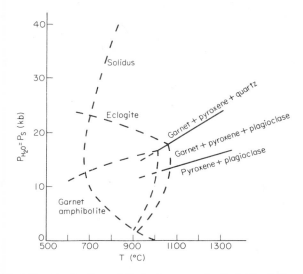

Fig. 4.17 Stability limit of amphibole in a quartz tholeiite bulk composition under conditions of $P_{H2O} = P_{total}$, compared with the basalt $\rightleftharpoons$ eclogite transition in dry quartz tholeiite composition. Note the negative slopes and relatively low melting temperatures in the water-saturated system. The generalised sub-solidus reaction of amphibolite to eclogite (a sliding reaction) is: hornblende$_A$+garnet$_B$+pyroxene$_C$ $\rightleftharpoons$ hornblende$_D$+garnet$_E$ + pyroxene$_F$ $\pm$ quartz $\pm$ olivine+water. After Essene, Hensen and Green (1970). *Physics Earth Planet. Interiors*, 3, p. 381.

struct various P-T or T-X grids showing the phase topology. However, a lot of the resulting geometry may be irrelevant if the reactions concerned never take place in rocks.

Similarly, many experiments concern simple reactions that define the stability limits of a phase or simple combination of phases. They may be reasonably applicable to simple rocks, but in the much more common complex situations they may tell us nothing about the P-T-X- conditions of the real prevailing reactions. However, they do give us P-T-X limits of phase stability (i.e. various assemblages in rocks can occur only within the stability limits of the phases concerned).

Also, there is not much point in studying the kinetics of specific reactions if they are irrelevant. So, a pressing need exists for the accurate delineation of actual reactions and their experimental investigation. This is most easily done in retrograde metamorphic situations, as discussed below.

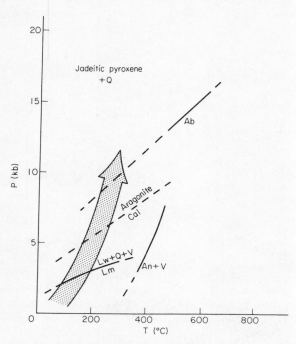

Fig. 4.18 P-T grid of main reactions relevant to the glaucophane-lawsonite schist facies, with inferred broad metamorphic gradient for this type of metamorphism. After Ernst (1971). *Amer. J. Science*, **270**, p. 98. Ab=albite; An=anorthite; Cal=calcite; Lm=laumontite; Lw=lawsonite; Q=quartz; V=water vapour.

Retrograde Reactions

Retrograde reactions are worth investigating, because commonly they (i) preserve original microstructures (i.e. they commonly are pseudo-morphous), (ii) involve only one of the original minerals in the rock, and (iii) do not go to completion, either for lack of reactants (principally water) or for kinetic reasons (Chapter 3). Therefore, they enable us to observe solid reactants and products, and so to infer realistic reactions.

For example, at Broken Hill, Australia, pegmatites have undergone deformation and partial retrograde metamorphism in retrograde schist-zones, along with other rock-types.[28,33,35] Replacement veinlets in

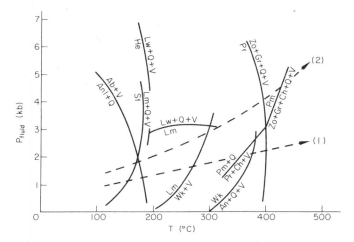

Fig. 4.19   P-T grid of experimentally determined reactions relevant to low-temperature regional metamorphism in the zeolite and prehnite-pumpellyite metagreywacke facies. Inferred P-T gradients are shown for metamorphism in (1) the Tanzawa Mountains, Japan and (2) the Taringatura area, New Zealand. After Liou (1971). *Contribs. Mineralogy and Petrology*, **31**, p. 175. Ab=albite; An=anorthite; Anl=analcite; Ch=chlorite; Gr=grossular; He=heulandite; Lm=laumontite; Lw=lawsonite; Pm=pumpellyite; Pr= prehnite; Q=quartz; St=stilbite; V=water vapour; Wk=wairakite; Zo= zoisite.

coarse alkali feldspar of fine-grained muscovite are fringed by lobes of myrmekite projecting into the remnant feldspar, so that the following reaction can be inferred as a reasonable approximation:[28]

$$\begin{Bmatrix} 3x\ KAlSi_3O_8 \\ y\ \ NaAlSi_3O_8 \\ zCa\ \square\ (AlSi_3O_8)_2 \end{Bmatrix} +2H^+ \ \rightleftharpoons$$

alkali feldspar

$$xKAl_2(AlSi_3)O_{10}(OH)_2 + \begin{Bmatrix} y\ NaAlSi_3O_8 \\ z\ CaAl_2Si_2O_8 \end{Bmatrix} + (6x+4z)SiO_2 + 2xK^+$$

muscovite   plagioclase   quartz

myrmekite

(where $\square$=cation vacancy, and $x>y\geqslant z$). Note the similarity to reaction (2) on page 32 (Chapter 2).

Similarly, in part of the Arunta Complex, central Australia, partial replacement of relatively coarse cordierite by acicular to vermicular aggregates of fine-grained kyanite, aluminous anthophyllite and quartz suggests that the following reaction may have taken place (based on electron microprobe analysis of the actual cordierite):[34]

$$5Mg_{1.74}Fe_{0.20}Al_{4.02}Si_{5.01}O_{18.00} + 2H_2O \rightleftharpoons 6Al_2SiO_5 + 7SiO_2$$

cordierite   kyanite   quartz

$$+ 2Mg_{4.35}Fe_{0.50}Al_{4.05}Si_{6.03}O_{22.00}(OH)_{2.00}$$

aluminous anthophyllite

Experimental work, using the natural minerals, suggests that the reaction may have proceeded in response to reintroduction of water after granulite facies metamorphism, rather than a change in P-T conditions.[13] So, inasmuch as 'grade' reflects a predominant change in temperature, this reaction may not be technically retrograde.

Although retrograde reactions generally are more amenable to investigation than most prograde reactions (which generally go to completion, thereby obliterating evidence of their mechanisms), they may not give us much insight into the details of prograde reactions, because (i) severe overstepping is possible, leading to metastable reactions in the retrograde situation, and (ii) commonly (though not always) only local systems react in retrograde metamorphism, whereas the whole assemblage commonly appears to react in most prograde metamorphism.

Also, we must be careful of deceptively simple retrograde reactions. For example, in the serpentinisation of alpine peridotites, pseudomorphs rich in serpentine commonly preserve the shapes of former olivine or orthopyroxene grains. From this we can infer reactions between olivine (or orthopyroxene) and water to give serpentine and minor by-products. But this is not the whole story, because although the former volume of the olivine grain is maintained, the volume of serpentine formed exceeds this, and so the excess serpentine must have migrated elsewhere (e.g. into chrysotile veins).

Similarly, other incomplete pseudomorphous retrograde reactions may be deceptively simple; for example, biotite or garnet partly pseudomorphed or veined by chlorite, and cordierite partly replaced by various lower-grade assemblages. Chemical considerations commonly show that other phases in the rock must have been involved. For example, though an assemblage such as chlorite + andalusite + quartz could have replaced cordierite without involving any other phase (except water), an assemblage such as chlorite + white mica + quartz could not replace cordierite unless other minerals were involved also, thus making the reaction more complex than it would otherwise appear; for example:

biotite + cordierite + water $\rightleftharpoons$ muscovite + chlorite + quartz

Reactions Involving Zoned Minerals

Here we have an interesting example of the essentially human nature of science. Before the advent of the electron probe microanalyser, chemical zoning in mineral grains could not be analysed, and so it had to be ignored, except where the approximate composition could be inferred optically. For example, Binns inferred that sodic cores in reversely zoned plagioclase in high-grade rocks at Broken Hill, Australia, represented metastable relics preserved from lower grades.[2] Therefore, minerals had to be separated mechanically for chemical analysis, so that the scale of equilibrium, of necessity, was the volume of the specimen from which the minerals were separated. Most metamorphic minerals were assumed to be relatively homogeneous, and, in fact, the supposed general absence of zoning was used as a point of evidence in favour of the general attainment of equilibrium in metamorphism.

Of course, the inferred absence of zoning was understandable for isotropic or colourless phases without optical evidence of zoning (e.g. many garnet grains), but the surprising (or not so surprising) thing is that, since the widespread availability of the electron microprobe, people are finding optically observable zoning in metamorphic minerals in many places. Whereas previously reactions were inferred from chemical analyses that averaged and so obscured any zonal variations, now we can use zoning to infer more realistic reactions.

An example of inferred reactions involving zoned minerals concerns the variation in the composition of zoned plagioclase and epidote in the Oak Lake—Whetstone Lake area, Ontario, described by Rambaldi.[29] Here the assemblage plagioclase-epidote occurs over a wide range of metamorphic grade in rocks of different bulk composition. The general increase in the anorthite content of plagioclase and concomitant decrease in the amount of epidote with increasing grade of metamorphism is well known, but the general reaction, Al-epidote $\rightleftharpoons$ anorthite + H_2O, is unbalanced, so that other minerals must participate in the actual reactions.

Rambaldi analysed the epidote and plagioclase with an electron probe microanalyser, and found that (i) some epidote grains are zoned in Fe^{3+} and Al, the epidote becoming progressively richer in Fe^{3+} with growth; (ii) there is no relationship between epidote composition and metamorphic grade; (iii) plagioclase grains are generally zoned, the zoning being well developed, commonly with two or three zones per grain;

(iv) in epidote-free rocks, the plagioclase is *albitic*, showing that (*a*) albite is stable in epidote-free assemblages throughout the amphibolite facies, and (*b*) the breakdown of epidote causes the increasing anorthite content of plagioclase.

A simple, general interpretation of these relationships is that, because plagioclase does not accept more than about 0.5% Fe_2O_3, it preferentially absorbs Al, leaving the epidote progressively enriched in Fe^{3+}. Thus, plagioclase showing reverse zoning is truly 'refractory', in the sense of Hollister (see below),[18] and so is the associated epidote where it shows outwards zoning in Fe^{3+}. However, whereas all the plagioclase is zoned, zoning in epidote is much less common, in which case the epidote is behaving as a 'matrix' mineral, according to Hollister.[18] As Rambaldi noted, homogenisation is more likely to occur in epidote (where it involves a simple exchange of Fe^{3+} and Al^{3+}) than in plagioclase (in which the migration of Na and Ca ions must occur in close association with a redistribution of Si and Al).[29] However, the plagioclase zoning is not always simply reversed, but may be normal and oscillatory, implying complexities in the actual reactions.

Rambaldi investigated many inferred reactions involving plagioclase and epidote. As an example, consider the following:

$$\text{actinolite} + \text{Al epidote} + \text{ilmenite} + \text{quartz} \rightleftharpoons$$
$$\text{hornblende} + \text{anorthite} + \text{sphene} + H_2O$$

The microstructural evidence for this reaction is: (i) the observed assemblage consists of amphibole, epidote, plagioclase and sphene; (ii) the plagioclase shows reverse zoning; (iii) the epidote is zoned with Fe^{3+} more abundant towards the rim; (iv) the amphibole consists of light green or colourless actinolite rimmed or almost completely replaced by dark green hornblende, which also forms small grains in the matrix; and (v) the sphene rarely encloses residual ilmenite.[29] Quartz is no longer present, but is inferred by the equation balance. So we see how microstructural and chemical data can be integrated to improve an understanding of a realistic reaction.

Reactions Involving Zoned Garnet

As mentioned previously, older studies ignoring garnet zoning are understandable. This appears to be reasonable for many granulites, peridotites and eclogites, in which the garnet is relatively homogeneous. But, since changes in the cation distribution coefficients between garnet and other phases are being increasingly used as P-T indicators, it is essential not to ignore zoning in most rocks. For example, Mn variations within one grain may be similar to those inferred as 'grade' variations by some workers. This needs explanation.

Hollister's study of contact metamorphism in the Kwoiek area of British Columbia is another example of the importance of microprobe analyses in the detailed inference of metamorphic reactions.[18] The area shows only one prograde metamorphic event, with no postmetamorphic deformation, and is a deep-seated contact metamorphic screen or roof pendant on the eastern edge of the Coast Range batholithic complex. The grade variation is from below a biotite isograd up to rocks that appear to have reached 700°C, over a distance of only four miles. The original rocks were greywackes and rocks of basaltic composition.

Garnet appears before staurolite, and Hollister inferred (from observation of assemblages) the following reaction:

$$\text{chlorite} + \text{muscovite} + \text{quartz} \rightleftharpoons \text{garnet} + \text{biotite} \\ + H_2O \qquad (1)$$

The appearance, at slightly higher grade, of staurolite in the assemblage involves the probable reaction (Fig. 4.20):

$$\text{chlorite} + \text{muscovite} + \text{ilmenite} \rightleftharpoons \text{staurolite} + \text{garnet} + \text{biotite} \\ + \text{quartz} + H_2O \qquad (2)$$

But the situation is more complicated than these simple reactions would suggest. The garnet and staurolite are zoned, whereas the biotite, chlorite and muscovite are unzoned, as indicated by electron microprobe traverses. Hollister's interpretation is that, as reactions proceed, the zoned ('refractory') phases store components that preferentially substitute in them (e.g. Mn in garnet), thereby depleting the surrounding matrix in those components. This causes outer zones to become progressively depleted in these components, whereas unzoned ('matrix') phases are able to equilibrate continuously to the changing chemical composition of the local system (Fig. 4.20). As a result, we are no longer dealing with a simple reaction involving homogeneous phases, but with a continuously changing distribution of chemical elements.

An observed Mn reversal and other changes near the edges of zoned garnet grains are interpreted by Hollister as being due principally to a change in the rock composition outside the garnet by a release of the components concerned into the system from a previously unreactive phase as it decomposed in a new reaction. Because reaction (1) runs at lower grades than reaction (2), most of the garnet had crystallised before reaction (2) occurred. The ilmenite remained unreactive until reaction (2), at which stage it liberated Ti that entered biotite and staurolite (as shown by microprobe analysis), and some Mn, which preferentially entered garnet and accounts for an observed Mn rise at the rim (i.e. an inflection in the compositional zoning curve).

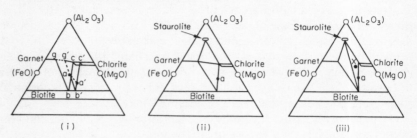

Fig. 4.20   AFM diagrams[31] to show phase and compositional relationships in some metapelitic rocks of the Kwoiek area, British Columbia.
(i) Because garnet stores Fe relative to Mg and the interior of the garnet is inferred to be removed from further reaction, the bulk composition of the 'effective system' (i.e. the matrix surrounding the garnet) is given by point a, and moves to a' as the reaction: muscovite+chlorite+quartz ⇌ biotite+garnet+H_2O proceeds. At the same time, the garnet-chlorite-biotite field moves to g'c'b', the biotite: chlorite ratio increases and the Mg:Fe ratios of biotite and chlorite increase.
(ii) For a bulk composition (a) below the garnet-chlorite join, garnet initially grows by the above reaction, as staurolite becomes a stable phase.
(iii) The reaction: chlorite+muscovite ⇌ staurolite+biotite+garnet+quartz +H_2O is illustrated by the replacement of the garnet-chlorite join (tie-line) by a staurolite-biotite join. So, bulk composition x should produce only staurolite-biotite-chlorite assemblages (+muscovite+quartz). The presence in the actual rocks of garnet as well has been attributed to metastable persistence caused by its failure to react with surrounding minerals. After Hollister, pp. 2489, 2491.[18]

Electron microprobe profiles for Mg, Fe and Mn of biotite and chlorite, on the other hand, show no regular variations and approach, though they do not attain, homogeneity. So, it appears that biotite and chlorite were able to equilibrate internally in response to changing conditions, in contrast to the garnet and staurolite. Also, the Mg/Fe ratio varies systematically from grain to grain. Hollister interpreted this as reflecting a gradual decrease in the volume of minerals in equilibrium with each biotite and chlorite grain, because of declining reaction rates during retrograde metamorphism.[18] Hollister noted that this suggested mechanism for garnet zoning would be most likely to apply to contact metamorphic environments, owing to a relatively short heating time, whereas garnet possibly may not be refractory in regional terrains.

Kretz, in a detailed interpretation, employed a similar general growth hypothesis to explain zoned garnet in low-pressure metapelitic schists from Yellowknife, Canada.[22] He explained the increase in Mn at the rims of the garnet grains by restriction of Mn diffusion in adjacent chlorite, so that the growing garnet grain consumed increasingly Mn-rich chlorite. He noted that a decrease in growth rate of the garnet might bring about the same result. Kretz also suggested that unzoned high-grade metamorphic garnet grains may have been zoned originally, but were homogenised by diffusion under prolonged high temperatures.

Anderson and Buckley's alternative explanation for garnet zoning is that it developed after the garnet grains had finished growing, on the assumption that garnet growth rates are fast relative to diffusion of the cations concerned in the garnet.[1] The idea is that the garnet exchanged components with the matrix by means of diffusion through the garnet lattice, which contrasts with the above-mentioned explanations based on growth without diffusion in the garnet. Inflections in the shapes of compositional zoning curves (such as those observed by Hollister and others) are to be expected from calculations based on the assumption that the diffusion coefficient for the element concerned is concentration- and time-dependent (i.e. if Fick's Second Law applies (Chapter 3) as found in many metal, oxide and glass diffusion experiments). Furthermore, if the matrix itself exchanged elements with adjacent volumes of 'local equilibrium', then the garnet-matrix exchange would be altered, producing inflections in the zoning compositional curve. Anderson and Buckley's model (which can be extended, in principle, to include growing garnet grains) is supported by the relative uniformity of reputedly immobile components (especially Si^{4+}, Al^{3+} and Ti^{4+}) in zoned garnet grains; any zoning of these components would be strong evidence in favour of compositional changes during growth of the garnet without diffusion in the garnet lattice.

Compositional Changes During Metamorphism ('Metamorphic Differentiation')

Interchange of elements along chemical potential gradients* may occur during metamorphism between (i) layers or pods of rock of differing

---

*The chemical potential (μ_i) of a component (i) in a system of variable composition is $\partial G/\partial n_i$ at constant P, T provided the molar concentrations of other components stay constant; where G = Gibbs free energy (Chapter 2) and n = number of moles of i. At constant P and T, $dG = \Sigma_i \mu_i dn_i$. If a small amount (dn_i) of i moves from mineral A to mineral B (or fluid phase B), the change in Gibbs free energy $(dG) = - \overset{A}{\mu}_i dn + \overset{B}{\mu}_i dn$. For this transfer to pro-

bulk composition, or (ii) individual grains of compositionally different minerals, to produce new compositional zones between them. Such gradients may be initiated or intensified by deformation, as discussed in Chapter 8, but they also occur in rocks showing no evidence of deformation at the time of the inferred gradients. For example, reaction between carbonate-rich and carbonate-free metapelitic rocks under open-system conditions can produce amphibolite, this being an effective mechanism for the formation of thinly layered amphibolites that previously were thought to form only by metamorphic differentiation of originally massive igneous rocks.[27] The suggested mechanism has experimental support, in that similar metasomatic zones occur in the laboratory investigation of carbonate-bearing and mica-bearing assemblages in juxtaposition.[36]

Differential transfer of K, Ca, Mg and probably Al occurred in these experiments, taking place almost entirely in the fluid phase, and being enhanced by higher temperatures, increased pore fluid and higher salt concentrations in the pore fluid. The transfer is believed to be initiated by a pre-existing compositional gradient, which, by differential transfer of elements, becomes a sequence of metamorphic mineral assemblages, each of which locally equilibrates with and buffers the pore fluid composition. Chemical species diffuse through the fluid down their own activity gradients and thereby change the composition of the metasomatic zones.[36]

Fisher has interpreted cores of andalusite + biotite + quartz surrounded by mantles of quartz + feldspar in a sillimanite-bearing metapelitic gneiss from Sweden, in terms of metamorphic segregation involving migration of K and (Fe, Mg, Ca) in opposite directions.[10] He suggested ionic equilibria forming exchange cycles linking migration of K *vs* Fe, K *vs* Mg and K *vs* Ca, such that K-feldspar (microcline) was depleted in the core and increased in the mantle. The suggested scheme for K *vs* Fe is shown below, chemical reactions being represented by single arrows, and migration of ionic components being represented by dashed arrows (see p. 131).

Fisher[11] has attempted to model diffusion-controlled processes such as this in terms of non-equilibrium thermodynamics, and has outlined

---

ceed spontaneously, dG < O, so that $\mu_i^A > \mu_i^B$. In other words, a component can move from phase A to phase B if its chemical potential is greater in A than in B (i.e. the addition of i to B lowers the free energy of the system). At equilibrium, dG = O, so that $\mu_i^A = \mu_i^B$; i.e. the chemical potential of i is equal in both phases, and this applies to all components in the two phases at equilibrium.[21, 32]

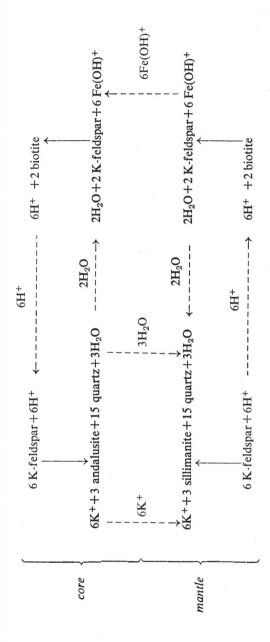

the theoretical conditions under which megascopic mineral segregations are likely to form.

Many other examples of reaction zones produced at the contacts of chemically dissimilar grains or aggregates of grains have been described in the literature.[24] Some examples of local metasomatic changes on the scale of a few grains have been mentioned in connection with staurolite-breakdown reactions and mafic corona reactions (see above). The zones or coronas reflect the varying stabilities of certain minerals or assemblages under varying chemical potentials of the mobile components, so that minerals in the reaction zones may be out of equilibrium with those in the immediately surrounding rock, even on the scale of a single thin section.[24] This further emphasises the concept of local systems (Chapter 2) and the tendency towards the establishment of local equilibrium in many metamorphic situations.

References

1 Anderson, D. E. and Buckley, G. R. (1973). Zoning in garnets—diffusion models. *Contribs. Mineralogy & Petrology*, **40**, 87–104.

2 Binns, R. A. (1964). Zones of progressive regional metamorphism in the Willyama Complex, Broken Hill district, New South Wales. *J. Geol. Soc. Australia*, **11**, 283–330.

3 Brown, G. C. and Fyfe, W. S. (1971). Kyanite-andalusite equilibrium. *Contribs. Mineralogy & Petrology*, **33**, 227–31.

4 Carswell, D. A. (1968). Possible primary upper mantle peridotite in Norwegian basal gneiss. *Lithos*, **1**, 322–55.

5 Carmichael, D. M. (1969). On the mechanism of prograde metamorphic reactions in quartz-bearing pelitic rocks. *Contribs. Mineralogy & Petrology*, **20**, 244–67.

6 Chatterjee, N. D. (1970). Synthesis and upper stability of paragonite. *Contribs. Minerology & Petrology*, **27**, 244–57.

7 Chatterjee, N. D. (1971). The upper stability limit of the assemblage paragonite + quartz and its natural occurrences. *Contribs. Mineralogy & Petrology*, **34**, 288–303.

8 Evans, B. W. (1965). Application of a reaction-rate method to the breakdown equilibria of muscovite and muscovite plus quartz. *Amer. J. Science*, **263**, 647–67.

9 Evans, B. W. and Guidotti, C. V. (1966). The sillimanite-potash feldspar isograd in western Maine, USA. *Contribs. Mineralogy & Petrology*, **12**, 25–62.

10 Fisher, G. W. (1970). The application of ionic equilibria to metamorphic differentiation: an example. *Contribs. Mineralogy & Petrology*, **29**, 91–103.

11 Fisher, G. W. (1973). Nonequilibrium thermodynamics as a model for diffusion-controlled metamorphic processes. *Amer. J. Science*, **273**, 897–924.

12 Gordon, T. M. (1971). Some observations on the formation of wollastonite from calcite and quartz. *Canadian J. Earth Sciences*, **8**, 844–51.

13 Green, T. H. and Vernon, R. H. (1974). Cordierite breakdown under high pressure, hydrous conditions. *Contribs. Mineralogy & Petrology*, **46**, 215–26.

14 Greenwood, H. J. (1972). Al^{IV}-Si^{IV} disorder in sillimanite and its effect on phase relations of the aluminium silicate minerals. *Geol. Soc. America Mem.*, **132**, 553–71.

15 Griffin, W. L. and Heier, K. S. (1973). Petrological implications of some corona structures. *Lithos*, **6**, 315–35.

16 Hobbs, B. E. (1968). Recrystallization of single crystals of quartz. *Tectonophysics*, **6**, 353–401.

17 Holdaway, M. J. (1971). Stability of andalusite and the aluminium silicate phase diagram. *Amer. J. Science*, **271**, 97–131.

18 Hollister, L. S. (1969). Contact metamorphism in the Kwoiek area of British Columbia: an end member of the metamorphic process. *Bull. Geol. Soc. America*, **80**, 2465–94.

19 Jones, J. W. (1972). An almandine garnet isograd in the Rogers Pass area, British Columbia: the nature of the reaction and an estimation of the physical conditions during its formation. *Contribs. Mineralogy & Petrology*, **37**, 291–306.

20 Kooy, G. (1970). Material transport in solid-state reactions in *Chemical and Mechanical Behavior of Inorganic Materials*, ed. Searcy, A. W. *et al.* 273–84. New York: Wiley-Interscience.

21 Krauskopf, K. B. (1967). *Introduction to Geochemistry.* 694–5. New York: McGraw-Hill Book Co.

22 Kretz, R. (1973). Kinetics of the crystallization of garnet at two localities near Yellowknife. *Canadian J. Earth Sci.*, **12**, 1–20.

23 Kridelbaugh, S. J. (1973). The kinetics of the reaction: calcite + quartz = wollastonite + carbon dioxide at elevated temperatures and pressures. *Amer. J. Science*, **273**, 757–77.

24 Kwak, T. A. P. (1974). Natural staurolite breakdown reactions at moderate to high pressures. *Contribs. Mineralogy & Petrology*, **44**, 57–80.

25 Mather, J. D. (1970). The biotite isograd and the lower greenschist facies in the Dalradian rocks of Scotland. *J. Petrology*, **11**, 253–75.

26 Miller, C. (1974). Reaction rims between olivine and plagioclase in meta-peridotites, Ötztal Alps, Bavaria. *Contribs. Mineralogy & Petrology*, **43**, 333–42.

27 Orville, P. M. (1969). A model for metamorphic differentiation origin of thin-layered amphibolites. *Amer. J. Science*, **267**, 64–86.

28 Phillips, E. R., Ransom, D. M. and Vernon, R. H. (1972). Myrmekite and muscovite developed by retrograde metamorphism at Broken Hill, New South Wales. *Miner. Mag.*, **38**, 570–8.

29 Rambaldi, E. R. (1973). Variation in the composition of plagioclase and epidote in some metamorphic rocks near Bancroft, Ontario. *Canadian J. Earth Sciences*, **10**, 852–68.

30 Strens, R. G. J. (1968). Stability of Al_2SiO_5 solid solutions. *Miner. Mag.*, **36**, 839–49.

31 Turner, F. J. (1968). *Metamorphic Petrology. Mineralogical and Field Aspects.* New York: McGraw-Hill Book Co.

32 Verhoogen, J., Turner, F. J., Weiss, L. E., Wahrhaftig, C. and Fyfe, W. S.

(1970). *The Earth. An Introduction to Physical Geology.* New York: Holt, Rinehart & Winston, Inc.

33 Vernon, R. H. (1969). The Willyama Complex, Broken Hill area. *J. Geol. Soc. Australia,* **16**, 20–55.

34 Vernon, R. H. (1972). Reactions involving hydration of cordierite and hypersthene. *Contribs. Mineralogy & Petrology,* **35**, 125–37.

35 Vernon, R. H. and Ransom, D. M. (1971). Retrograde schists of the amphibolite facies at Broken Hill, New South Wales. *J. Geol. Soc. Australia,* **18**, 267–77.

36 Vidale, R. (1969). Metasomatism in a chemical gradient and the formation of calc-silicate bands. *Amer. J. Science,* **267**, 857–74.

Chapter 5

Stable Metamorphic Microstructures

Introduction

When an unstable aggregate of minerals reacts in a metamorphic environment, new minerals are formed in order to reduce the *chemical free energy* of the rock system concerned. However, even after the rock space has become fully occupied by grains of the new, stable minerals and the chemical potential energy has been reduced to a minimum, further changes can occur. These involve changes in grain shape, in response to *interfacial free energy* (grain-boundary energy). This energy is much smaller than the chemical free energy, but it is effective nevertheless in controlling the shapes of grains in metamorphic rocks, the grain boundaries arranging themselves in configurations that reduce the grain-boundary potential energy to a minimum.

The reason that grain boundaries are associated with a free energy is that they are narrow zones (a few atoms wide) in which the atomic structure is much more disordered than inside the grains. Owing to the unlikelihood of perfect compromise between the lattices of neighbouring grains meeting at random, atoms in these zones where the lattices join together ('grain boundaries') are in positions of higher potential energy than they would be if they were accommodated in regular arrays inside grains. So, the tendency always is for an aggregate of grains to attempt to reduce the proportion of atoms in unstable grain-boundary sites (compared with those in more stable sites inside grains). This can be done either (*a*) by reducing the total area of the grain boundaries in an aggregate, or (*b*) by growing certain stable or metastable crystal faces in which the atoms occur in positions of much lower energy than they would if the grain boundary were non-crystallographic (random, irrational). In metamorphic rocks both processes operate to varying degrees, depending on the actual minerals present.[8, 17]

Growth and adjustment of crystalline grains in the solid state produce certain grain-shapes, provided enough time and heat are available for stability to be approached relatively closely.[1, 6, 7, 11, 12, 13] In this chapter,

I will assume that a stress-field is absent, or has a negligible effect, during this *grain growth*. The microstructures produced are common in many metamorphic rocks.[2, 8, 14, 15, 17, 20]

Unstable Grain Arrangements

Grain arrangements with high interfacial free energy are those with small grain sizes (large grain-boundary areas), and/or large porosities (large surface areas of crystalline material against fluid), and/or irregular grain shapes (large interfacial and/or surface areas). Examples are:

(i) accumulations of fragments in sediments, which can vary greatly in size, shape and degree of sorting, and which have a minimum porosity of around ten per cent, even after compaction of the finest-grained material;

(ii) fine-grained chemical precipitates (e.g. chert and chemical limestone);

(iii) strongly deformed crystals or grain aggregates (e.g. mylonites) in which recrystallisation has occurred to minimise strain energy (Chapters 6, 7), either by the development of strain-free nuclei or by the gradually increasing misorientation of subgrains; once the deformed grain has been replaced by a space-filling aggregate of strain-free grains, the aggregate can undergo further growth and adjustment to minimise the interfacial energy present as a result of fine grain size and/or relatively irregular grain shape;

(iv) a polyphase metamorphic aggregate that has stabilised (minimised chemical-free energy under equilibrium or steady-state conditions), so that a space-filling aggregate of chemically stable (or metastable) grains has been produced. Further growth and adjustment may occur to minimise interfacial energy present as a result of fine grain-size and irregular shape (due to fortuitous impingement of grains).

Stable Grain Arrangements in Isotropic Single Phase Aggregates

A general idea of the grain shapes that result from grain growth can be obtained from the examination of foams (e.g. soapy water-air and especially beer-air froths, which are displayed excellently in the necks of beer bottles after pouring beer from them—an experiment well worth attempting, in more ways than one*). Annealed metals, sintered ceramics, glacier ice and organic cellular aggregates all show similar grain or cell configurations.[1, 11, 12, 13] A typical two-dimensional aggregate

*All observations should be made early in the experiment.

looks like that shown in figure 5.1. Three grains (cells) meet at a point ('triple-junction'), the interfacial angles closely approximating 120°, these being the obvious geometrical results of an attempt to fill space and reduce the interface area to a minimum in an aggregate containing interfaces of equal free energy.[6] Large grains with many sides grow bigger at the expense of grains with few sides, which get smaller, eventually becoming three-sided (in two dimensions) before they disappear altogether.[1,6,11] Grain boundaries move towards their centres of curvature. In this way, the grain-size increases. The process slows down as interface curvatures decrease, but in foams all boundaries eventually disappear. In crystalline materials this stage is not reached, many boundaries remaining. Grain growth can also be slowed down by many small grains of a dispersed phase.[20] For example, in quartzites, layers containing many small mica grains have much finer-grained quartz than mica-free layers (Fig. 5.2).

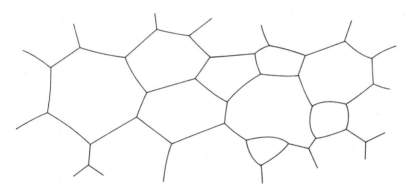

Fig. 5.1 Two-dimensional section through a typical polygonal aggregate that has attained a relatively high degree of interfacial stability in a structurally isotropic material.

Anisotropic Single Phase Aggregates

Foams and organic cellular aggregates are truly *isotropic*, so that interfaces are of equal energy and 120° angles are attained.[12] However, all crystalline materials (even cubic ones) are *anisotropic*, with regard to their atomic structures, and hence many of their physical properties.

This means that the interfacial free energy is also anisotropic, so that some grain boundaries have smaller free energies than others. Statistically, most boundaries probably can be regarded as high-energy boundaries, except for minerals that are *strongly anisotropic* (see below). The effect of minor anisotropy is to increase the spread of interfacial angles about the mean of 120°, and to cause local deflection in otherwise smoothly curved interfaces.[8, 11, 17] However, the aggregate is still polygonal and resembles foam. This is commonly the situation in high-grade, metamorphic aggregates of quartz, feldspar, garnet and calcite.[17] Polygonal aggregates of this

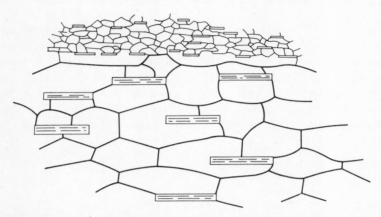

Fig. 5.2   Two-dimensional hypothetical section through a layered micaceous quartzite, suggesting that growth of quartz grains has been retarded by the presence of numerous small mica flakes in the top layer. Also note that quartz/quartz interfaces meet mica $_{001}$/quartz interfaces at approximately 90° or are joined to edges of the mica flakes.

type have been produced experimentally in ice, quartz, calcite, fluorite, anhydrite and sulphides.[2, 15] In somewhat more anisotropic minerals, a few interfaces may be partly parallel to planes of specific low-index crystal forms, such as {110} of hornblende or pyroxene, the rest being smoothly curved and irrational.[8, 17]

In still more anisotropic minerals (such as mica and sillimanite) low-index forms (e.g. {001} in mica and {110} in sillimanite) typically predominate in the aggregate, producing many boundaries of an *impingement* rather than *adjustment* type.[8, 17] 'Decussate' aggregates of biotite provide common examples in metamorphic rocks (Fig. 5.3). These planar, rational interfaces are stable or metastable.

Aggregates with More than One Phase

The same considerations apply to two-phase (Figs 5.4, 5.5) and polyphase aggregates, except that the angle formed where a grain of one phase (α) meets two grains of another phase (β) varies with the phases concerned. If the energy (γ) of the grain boundary (β/β) is greater than that of the interphase boundaries (α/β), the dihedral angle ('$\theta\alpha$ *vs* β/β') is less than 120°, and vice versa (Fig. 5.4).[11] In effect, the dihedral angles become adjusted so that as large a total area of lower-energy α/β boundary

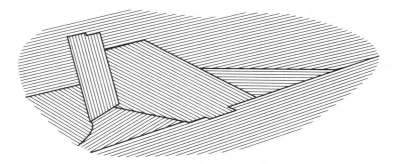

Fig. 5.3 'Decussate' aggregate of mica grains, in which most interfaces are parallel to {001} of one of the grains concerned. These boundaries may be referred to as 'rational-impingement' interfaces.

replaces higher energy β/β boundary as is appropriate to the energy differences involved. A mechanical analogy is commonly used so that some people refer to the higher interfacial 'tension' of a β/β boundary tending to 'pull in' the α/β boundaries of lower 'tension', thus forming

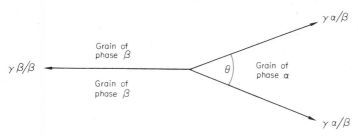

Fig. 5.4 Situation, in section, at the intersection of a grain of one phase (α) with two grains of a second phase (β), showing the operative interfacial energies or 'tensions' (γ) and the dihedral angle ($\theta\alpha$ *vs.* β/β).

a smaller dihedral angle. Many measurements have shown that generally, though not always, $\theta\alpha$ *vs* β/β is less than 120°.[11, 13, 15, 17] In other words, boundaries between grains of the same phase tend to have higher energies than those between grains of different phases.

In polyphase aggregates involving both anisotropic and effectively isotropic minerals, rational planes of the anisotropic minerals tend to dominate the microstructure. Common representatives in rocks are boundaries between (001) planes in layer silicates and, for example, quartz or feldspar. Thus, there is little tendency for a quartz/quartz or quartz/feldspar interface to alter the course of such a stable boundary. The result is that quartz/quartz (etc.) interfaces commonly meet mica$_{(001)}$/quartz interfaces at 90°, as though the latter interfaces remain 'inert' in the grain-adjustment process (Fig. 5.2).[20] However, in the third dimension, the quartz/quartz boundary, if curved, can move across the mica (001) plane and become attached to its edge with some other mica plane (Fig. 5.2), thereby tending to be anchored. In this way, layer silicate and similar grains tend to restrict grain growth in quartz and feldspar aggregates, and also cause elongate shapes governed by the length and separation of layer silicate grains (Fig. 5.2), which is a simple form of *mimetic* grain growth (i.e. growth controlled by pre-existing grain arrangements).

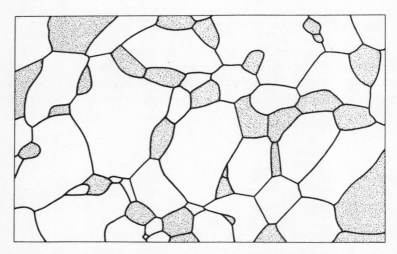

Fig. 5.5   Plagioclase (clear) - clinopyroxene (stippled) aggregate in a basic granofels showing polygonal grain-shapes. Magnification × 50. After Vernon, p. 339[18]. By permission of Scottish Academic Press Ltd.

Inclusions are sensitive indicators of anisotropy. The interface between an inclusion and its host is a true grain boundary, so that, even though the inclusion is a grain of the stable mineral assemblage, and though the diffusion rates within the host are too slow for the inclusion to be eliminated altogether, the tendency will be for the inclusion to reduce its boundary energy to a minimum. If both inclusion and host are isotropic, the resulting ideal shape will be a sphere, which has a minimum

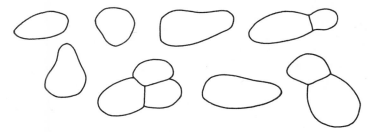

Fig. 5.6 Typical rounded shapes of inclusions of one effectively isotropic phase in another (e.g. quartz in K-feldspar). Note the dihedral angles formed at interface intersections in multiple inclusions.

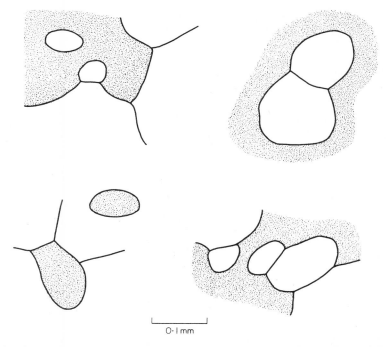

0·1 mm

Fig. 5.7 Grain and inclusion shapes in olivine (stippled) - plagioclase (clear) aggregates in allivalite adcumulate from the Rhum layered complex. After Vernon, p. 348.[18] By permission of Scottish Academic Press Ltd.

area per unit volume, and, if both minerals are not too anisotropic, approximately spherical or elliptical shapes are approached (Figs 5.6, 5.7). If one phase is 'effectively' isotropic and another strongly anisotropic, planar low-index boundaries may appear, these effectively lowering the grain-boundary free energy (Fig. 5.8). Even so, sites of high atomic misfit (and hence high potential energy), such as corners of polyhedral grains, are generally 'rounded off' (Fig. 5.8).[17]

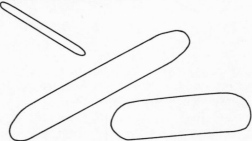

Fig. 5.8　Typical shapes of inclusions of a strongly anisotropic phase (e.g. apatite or sillimanite) in an effectively isotropic phase (e.g. quartz).

It must be re-emphasised that all the foregoing discussion applies only to random grains grown to a high degree of stability in the absence of a stress field. At low grades of metamorphism these 'ideal' shapes may or may not be achieved, and growth in a stress field can give rise to elongate shapes and strong preferred orientations.

Spatial and Size Distribution of Grains in Metamorphic Rocks

Some metamorphic rocks, especially those formed at lower grades and without deformation, show microstructural relics that permit reaction histories to be interpreted from growth sites of new minerals. This applies also to some retrograde metamorphic reactions (Chapter 4), which commonly are incomplete, allowing recognition of both the old and new grains.[19] However, in most metamorphic rocks, which have been extensively recrystallised, the original microstructure has been obliterated, so that interpretation of physical and chemical history on the grain scale is very difficult. Original, relatively large, compositional heterogeneities (such as bedding, earlier metamorphic foliation, etc.) may be retained, and 'metamorphic differentiation' (Chapter 8) may have intensified an earlier compositional heterogeneity or produced a new one. In any event, within a compositional unit (e.g. layer), the distribution and size of grains is determined by general controls, as yet incompletely understood.

For example, the following general observations of metamorphic microstructures need explanation:[3]

(i) a generally small range of variation in grain-size of a particular mineral in a rock;
(ii) a generally small range of variation in relative grain-sizes between different minerals in the same rock;
(iii) a relatively uniform periodic spatial distribution of various minerals—commonly through large volumes of rock, but also within small layers or pods of a mesoscopically heterogeneous rock.

As we have seen, an aggregate of grains can reduce its total interfacial free energy by (*a*) forming some interfaces in low-energy orientations, and (*b*) by forming grains as large as possible, thereby reducing the total interface area. However, in addition, we have seen that, in two-phase aggregates, boundaries between grains of two phases commonly have lower free energies than boundaries between grains of the same phase. So, polyphase aggregates conceivably could reduce their total interfacial free energy by forming as many *interphase* boundaries as possible.

The distribution of grains in quartzofeldspathic metamorphic rocks, although apparently random,[10] can be interpreted by certain statistical techniques as a slight departure from randomness.[5] Boundaries between grains of the same phase occur less commonly than expected for a random distribution. Furthermore, an increase in the proportion of interphase boundaries with increasing metamorphic grade (i.e. inferred increasing metamorphic stability) has been observed in one well-studied area of quartzofeldspathic rock.[4]

The mechanisms, by which this comes about, are understood only in broad outline.[3, 16] One suggestion is that a new phase nucleates and grows in such a way that high-energy interfaces or parts of interfaces are replaced first.[3] Succeeding grains are forced to occupy increasingly less energetically favourable positions in the aggregate, so that the total interfacial energy could decrease initially, followed by a steady situation. After this stage, the interfacial energy may even increase if the same grains continue to grow, as grains are brought more into contact with grains of unfavourable orientation and/or composition. Such a situation may force the nucleation of new grains and/or the preferential growth of existing grains of the phase concerned at other more favourable sites (that may have developed during the grain growth and adjustment). So, the general process may involve a constant shifting of crystal growth from one site to another, and back again, as the interfacial energy situation undergoes local changes.[3] These growth limits probably contribute greatly to the above-mentioned relatively even distribution and grain-size in many simple metamorphic polyphase aggregates. If this mechanism is generally correct, it follows that any solid-state process by which many boundaries between grains of the same phase are produced (such as in monomineralic layers produced or intensified by 'metamorphic differentiation') must be driven by some source of energy other than interfacial free energy.

Another way an aggregate can reduce its interfacial energy is by adsorption of 'impurity' ions in the grain and interphase boundaries, but this effect has not been studied quantitatively for minerals (except for some very preliminary experiments showing that mica has a lower surface energy in air than in a vacuum).[8] However, it merits detailed investigation, as it could affect interfacial angles and the areas of particular interphase boundaries (especially if adsorption occurs preferentially on them).

Growth of isolated phases in solid aggregates must occur in the early stages of metamorphism, and if their nucleation rate ($\dot{N}$) is small, these phases may remain isolated, commonly as grains ('porphyroblasts') much larger than the grain-size of the remainder of the rock. The size and distribution of such isolated grains is not controlled predominantly by interfacial free energy (although their final shapes may be), but by the rate of nucleation ($\dot{N}$) relative to the rate of growth ($\dot{G}$). If the ratio $\dot{N}:\dot{G}$ is large, many small grains are produced, and vice versa. In solid-state laboratory experiments, $\dot{G}$ commonly has been found to be constant for isolated grains of a particular phase, so that the grain-size distribution would mainly reflect $\dot{N}$, especially for the earlier stages of growth.[9] However, growth discontinuities (marked by compositional zoning and/or crystallographically arranged zones of inclusions) are common in natural porphyroblasts, so that the assumption is not strictly true, especially in the declining periods of growth. In fact, discontinuities occur in the grain-size frequency distributions of all metamorphic minerals investigated so far, even skewed bimodal distributions having been observed. These discontinuities are probably due to fluctuations in either $\dot{N}$ or $\dot{G}$, owing to temperature variations, local diffusion constraints, etc.[9, 14]

The spatial distribution of garnet in a metapelitic rock has been found to be random, and the size of a particular garnet grain is unrelated to the distance to its nearest neighbour. Therefore, we cannot say that all nuclei of a phase are formed at the same time in metamorphic rocks, if $\dot{G}$ is relatively constant.[9]

The shapes of porphyroblasts vary from xenoblastic and irregular (e.g. cordierite in many hornfelses) to idioblastic (e.g. staurolite in many schists). Crystal faces on porphyroblasts are hard to explain by normal mechanisms of grain growth. The relatively sudden development of grains much larger than their neighbours is well known in annealed metals and ceramic materials,[1, 7] the phenomenon probably being due to the fortuitous coalescence of one or more grains that happen to be in such an orientation that boundaries between them disappear. Once initiated, these large grains grow rapidly, because they have many more sides than the other grains in the aggregate, the process being known as *secondary recrystallisation* or *exaggerated grain growth*. The final porphyroblast should show concave-outwards cusps along its boundary, reflecting adjustment to neighbouring grains. However, many porphyroblasts (in metals and ceramics, as well as rocks) appear to have perfectly planar faces, in the light microscope at least. This has been explained by postulating a thin layer of fluid phase around the porphyroblast, so that actually it is growing in a fluid, and so can adopt a minimum energy crystal shape independent of grain neighbours. I wonder whether the same result might be achieved by the accumulation along the boundary of material ('impurity') unwanted by the growing porphyroblast, this impurity altering the interfacial free energy to such an extent that crystal faces can develop (Chapter 2).

The proportion of inclusions in a porphyroblast depends on the rate of growth of the porphyroblast relative to the rate of diffusion of included material through the porphyroblast and along its boundaries. Once a grain has been completely incorporated in another grain, it appears to be difficult to remove, presumably because of the relative slowness of diffusion through silicate lattices under most metamorphic conditions. However, if a boundary of a porphyroblast moves slowly enough and if the rates of grain-boundary diffusion are high enough, the material of the small grain may diffuse along the grain boundary to join some larger grain of the same phase. However, especially in the early stages of metamorphic grain development, the grain boundaries may move too fast (relative to grain-boundary diffusion rates) for this to occur. The result is a '*poikiloblastic porphyroblast*' (e.g. cordierite in many fine-grained hornfelses), containing numerous inclusions. Commonly inclusions in the central parts of porphyroblasts are smaller than the grain-sizes of the same phases in the aggregate outside the porphyroblast, suggesting that they were incorporated at a relatively early stage of grain coarsening in the rock.

Environments of Grain Adjustment in the Solid State

Though the foregoing discussion refers especially to crustal metamorphic rocks, it applies equally to aggregates in the earth's mantle (as inferred from fragments of peridotite and eclogite brought up in basaltic magmas) and to slowly cooled igneous rocks.[18,20] However, in igneous rocks a relatively high proportion of rational (low-energy) boundaries may be present. These tend to be stable, so that even though some grain boundary adjustment of space-filling aggregates (e.g. plagioclase and pyroxene in gabbro) could occur during slow cooling, the typically metamorphic polygonal (granoblastic) microstructure generally is not achieved. Accumulates commonly show a strong tendency towards polygonal microstructures and adjustment of inclusion shapes (Fig. 5.7), presumably owing to adjustment of grain boundaries in the absence of a fluid phase (i.e. after impingement of grains had occurred).[18]

References

1 Brophy, J. H., Rose, M. and Wulff, J. (1964). *The Structure and Properties of Materials.* Vol. II. *Thermodynamics of Structure*, 132–45. New York: J. Wiley and Sons, Inc.
2 Buerger, M. J. and Washken, E. (1947). Metamorphism of minerals. *Amer. Mineralogist*, **32**, 296–308.
3 De Vore, G. W. (1959). Role of minimum interfacial free energy in determining the macroscopic features of mineral assemblages. I. The model. *J. Geology*, **67**, 211–27.
4 Ehrlich, R., Vogel, T. A., Weinberg, B., Kamilli, D., Byerly, G. and Richter, H. (1972). Textural variation in petrogenetic analysis. *Bull. Geol. Soc. Amer.*, **83**, 665–76.
5 Flinn, D. (1969). Grain contacts in crystalline rocks. *Lithos*, **3**, 361–70.
6 Harker, D. and Parker, E. R. (1945). Grain shape and grain growth. *Trans. Amer. Soc. Metals.*, **34**, 159–95.
7 Kingery, W. D. (1960). *Introduction to Ceramics*, 191–216; 353–80; 402–17. New York: J. Wiley and Sons, Inc.
8 Kretz, R. (1966). Interpretation of the shape of mineral grains in metamorphic rocks. *J. Petrology*, **7**, 68–94.
9 Kretz, R. (1966). Grain-size distribution for certain metamorphic minerals in relation to nucleation and growth. *J. Geology*, **74**, 147–73.
10 Kretz, R. (1969). On the spatial distribution of crystals in rocks. *Lithos*, **2**, 39–66.
11 Smith, C. S. (1948). Grains, phases and interfaces: an interpretation of microstructure, *Trans. Amer. Inst. Min. Metall. Engineers*, **175**, 15–51.
12 Smith, C. S. (1954). The shape of things. *Scientific American*, **190**, 58–64.
13 Smith, C. S. (1964). Some elementary principles of polycrystalline microstructure. *Metallurgical Reviews*, **9**, 1–48.
14 Spry, A. (1969). *Metamorphic Textures.* Oxford: Pergamon Press.
15 Stanton, R. L. (1964). Mineral interfaces in stratiform ores. *Proc. Inst. Mining Metall.*, **74**, 45–79.
16 Verhoogen, J. (1948). Geological significance of surface tension. *J. Geology*, **56**, 210–17.

17 Vernon, R. H. (1968). Microstructures of high-grade metamorphic rocks at Broken Hill, Australia. *J. Petrology*, **9**, 1–22.

18 Vernon, R. H. (1970). Comparative grain-boundary studies of some basic and ultrabasic granulites, nodules and cumulates. *Scottish J. Geol.*, **6**, 337–51.

19 Vernon, R. H. (1972). Reactions involving hydration of cordierite and hypersthene. *Contribs. Mineralogy & Petrology*, **35**, 127–35.

20 Voll, G. (1960). New work on petrofabrics. *Liverpool and Manchester Geological Journal*, **2**, 503–67.

Chapter 6

Deformation, Recovery and Recrystallisation Processes

Introduction

This chapter is concerned with the general processes underlying the deformation, recovery and recrystallisation of crystalline materials. The general principles have been summarised for structural geologists in leading text-books.[9, 15] After a discussion of brittle *vs* ductile deformation, I will deal with general aspects of crystal plasticity, followed by a discussion of dislocations (line defects) in crystals and their effect on deformation, recovery and recrystallisation.

Brittle Versus Ductile Deformation

Geologically speaking, ductility is the capacity for undergoing permanent change of shape without gross fracturing.[11] It depends on the scale of observation. For example, a body of rock may appear to be undergoing deformation without gross fracturing when viewed on one scale (and so we would call it ductile), and yet the deformation may be heterogeneous when viewed on a finer scale. The heterogeneities may be slip (a form of plastic deformation) or fine-scale fracturing ('cataclasis' or brittle deformation). Since we are concerned with rather small-scale processes in this book, I will discuss plastic versus brittle deformation on the scale of a grain or small aggregate.

Crystal Plasticity

Because the crystalline state has the lowest possible free energy, solids attempt to remain crystalline when subjected to stresses tending to deform or fracture them. This has been proved many times by X-ray diffraction studies of strongly deformed materials. For example, sapphire (Al_2O_3) can be deformed up to a strain of 20 per cent without altering its

X-ray pattern significantly. Al and NaCl produce diffuse X-ray patterns after strong deformation, but remain essentially crystalline.

Primary Modes of Intragranular Plastic Deformation

In order to preserve crystallinity, only two fundamental modes of deformation of single crystals are possible (at temperatures too low to permit extensive diffusion and at pressures too high for brittle fracture), namely:

(*a*) *translation gliding* ('*slip*'), in which layers of the crystal glide over one another by amounts equal to the size of the unit pattern of the crystal structure. The pattern always comes back into registry afterwards (Fig. 6.1). This means that it cannot be detected optically in rock thin

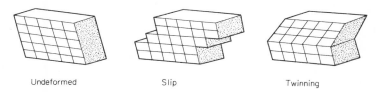

Undeformed   Slip   Twinning

Fig. 6.1 Diagrams showing the macroscopic changes involved in slip and mechanical twinning.

sections, unless some marker has been offset or rotated by the slip. Slip may occur in multiples but not fractions of the unit repeat distance. It usually occurs on widely separated planes; i.e. it is usually very heterogeneous. Slip planes are usually planes of relatively dense atomic packing, and slip directions are usually densely populated rows in these planes, in order to make atom jumps as small as possible. The combination of a slip plane and a slip direction in that plane is known as a *slip system* (e.g. (0001) plane and an *a*-axis in quartz). Methods for working these out in silicates are discussed at the end of this chapter.

(*b*) *twin gliding* (deformation twinning, mechanical twinning), in which each layer of the crystal structure is sheared (not translated) by an amount exactly sufficient to produce a mirror image of the original crystal (Fig. 6.1). Although the crystal structure is restored by the process, each half of the twin is misoriented symmetrically with respect to the other. Twinning is a homogeneous mode of deformation, inasmuch as the deformation is homogeneous on the scale of atomic planes within the twinned region. The distribution of twins within a grain, however, is generally very heterogeneous. Twinning tends to be favoured over slip at lower temperatures and faster strain rates. Although macroscopically twinning appears to involve simple shearing, detailed consideration of atomic structures commonly indicates either that (*a*) extra diffusional or non-diffusional movements ('shuffles') of atoms are required to bring the structure back into correct registry, or (*b*) the twinning can be achieved by a different displacement to that suggested by macroscopic considerations alone.

Secondary Modes of Deformation

These are derived from the primary modes and occur particularly when the grains are constrained during deformation, and so are forced to deform in various ways, such as bending, twisting and kinking (Fig. 6.2),

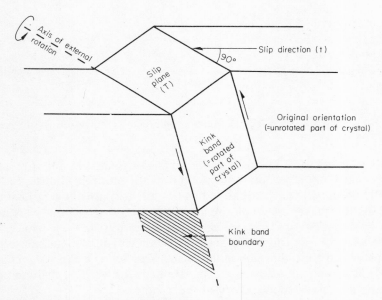

Fig. 6.2 Geometry of plastic kinking.

in order to change their shapes to conform to surrounding grains. These are all different functions of the simple slip process. For example, the relationship between plastic kinking and the slip system is shown in figure 6.2. If grains could not deform in these various ways, they would fracture along the grain boundaries, and the aggregate would deform by brittle, not ductile, deformation (on the scale of several grains). Because of the complex strain and stress distribution in an aggregate, only qualitative relationships exist between the mechanical behaviour of single crystals and polygrain aggregates.

For a polygrain aggregate to undergo a general strain by intragranular slip alone, without losing cohesion, it has been found that the material must have at least five independent slip systems (the von Mises criterion).[11] This is no problem for materials with high structural symmetry, such as cubic metals, which are characteristically ductile at relatively low temperatures. Polygrain NaCl (halite) is brittle at temperatures at which either only its $\{110\}<\bar{1}10>$ or $\{100\}<011>$ slip systems are active, but becomes very ductile at higher temperatures, at which both systems operate. Hexagonal close-packed metals and ice are very ductile as single crystals at certain temperatures, but their polygrain equivalents are much less ductile at the same temperatures, because slip on (0001), which characterises these materials, does not provide enough independent slip systems.

Concerning minerals, calcite and quartz have enough slip systems to satisfy the von Mises criterion, but no other known minerals have. Many (including most silicates, e.g. kyanite, enstatite, diopside, some feldspars) have only one or two systems. So, the common observation that calcite and quartz appear to be among the most ductile minerals in rocks is reasonable, as is the observation that feldspars and other minerals commonly appear to have behaved in a brittle manner under the same conditions. However, in some situations minerals appear to be very ductile, despite the deficiency of known slip systems. For example, polygrain talc (which only has known slip systems in its basal plane) is so weak that it is used as a medium for transmitting pressure in some high-pressure experimental equipment. Similarly, polygrain olivine, which has only three known slip systems, shows evidence of ductile behaviour under some natural conditions.

Therefore, some means of relaxing the von Mises restriction must apply under certain conditions. Some possibilities[11] are as follows:

(i) *Repeated planar concentrations of shearing strain*, such as mechanical twins or kinks, can be viewed as the approximate equivalent of an additional 'slip system', if they are repeated on a fine enough scale. However, both twinning and kinking are restricted, compared with slip, because they have only one sense of shear. Kinking involves simple shear parallel to the kink band boundary in a direction normal to the kink axis (intersection of kink band boundary and slip plane in the undeformed part of the grain), as shown in figure 6.2. This means that a fine set of kink bands is effectively equivalent to a new slip system independent of the primary slip system that produced the kink. Kinking appears to be an important contributor to the ductility of minerals with only one slip plane (e.g. talc, mica and kyanite) or of minerals with only a few slip systems (e.g. olivine, in which kinks, or deformation bands, are common). Kinking is also common in some minerals with sufficient slip systems to satisfy the von Mises criterion (e.g. quartz).

(ii) *Heterogeneous strain* could have an effect, since the von Mises criterion is based on an assumption that deformation is homogeneous on the grain scale, which it demonstrably is not. Most grains may approximate the macroscopic strain, but local departures from this may occur in places in the aggregate. However, the magnitude of this effect is not known.

(iii) *'Cataclasis'* may occur where the strain requirements cannot be met by slip, by either avoiding misfits between grains, or acting as an accommodation mechanism on an intragranular scale. For example, small accommodation fractures occur at the terminations of some experimentally produced deformation twins in plagioclase (Chapter 7), and natural large-angle kink-like structures in plagioclase have involved accommodation fracturing along the plane of misorientation (Chapter 7). Another example is the common formation of accommodation voids in larger-angle kinking in naturally and experimentally deformed mica (Paul Williams, personal communication). All these microfractures are local and do not propagate to cause gross fracturing of the rock, because the stresses are predominantly compressive in geological environments,[11] so that on a coarse enough scale the deformation is 'ductile'. Cataclasis is likely to be most important at lower temperatures and confining pressures, and both brittle and ductile intragranular deformation may accompany each other under the right conditions.

(iv) *'Diffusional creep'* can help to accommodate misfits caused by an insufficient number of slip systems, at temperatures high enough to promote adequate diffusion, and/or at slow strain rates. Dislocation climb (see later) and possibly general multidirectional diffusion of point defects can minimise the limitations on ductility. 'Grain boundary sliding' (see later) can occur also at higher temperatures, contributing to the strain, and being limited by the rate at which grains can accommodate to it by intragranular deformation processes at grain corners. But, the process apparently does not relax the von Mises limitation.[11]

Dislocations[3, 5, 7, 10, 13, 17]

Point defects (discussed in Chapter 3) are thermodynamically stable, because the increase in free energy associated with elastic distortion at the defect is outweighed by the increase in entropy associated with the formation of the defect.

Dislocations (or line defects) are thermodynamically unstable, because the increase in entropy is outweighed by the increase in elastic strain energy along the line of the defect. Theoretically, all processes should tend towards the production of a single homogeneous crystal with point defects only. However, in rock processes this rarely is achieved, for kinetic reasons.

Dislocations are produced during growth of crystals (by stacking errors, for example) or during deformation. Their recognition in 1934 by metallurgists led to a reconciliation between theory and experiment, which had been in violent disagreement. For example, crystals were observed to grow many times faster than predicted by normal nucleation and growth models, but if dislocations outcropping at the surface of the crystal assisted growth, the discrepancy disappeared. Similarly, metals were observed to be many times weaker than predicted theoretically, until dislocations were invoked to assist deformation. So the concept of dislocations is fundamental to an understanding of crystalline deformation and growth processes, and though the details of dislocations in minerals are just beginning to be known, their general nature (especially as revealed in simpler materials, such as metals) is well worth understanding.

How Do Dislocations Assist Deformation?

The simplest idea of slip is that two blocks of the crystal slide over each other (Fig. 6.1), but the friction would be prohibitive. Try pulling a large carpet over a floor! However, it is easy if a buckle is formed in the carpet and then the buckle is pushed forwards. The result is a net displacement of the whole carpet by an amount equal to the wavelength of the buckle, although at any one time only the small area of carpet under the buckle is not in contact with the floor. Similarly, in intracrystalline slip a dislocation (or series of dislocations) does the job by localising the strain, while still allowing a net displacement (Fig. 6.3). Apart from the local region of the dislocation itself, all atoms are linked normally. Thus dislocations greatly reduce the stress needed for translation gliding.

A dislocation line nucleates at a site of locally high stress, and moves through a crystal as glide occurs, being simply the boundary between translated and untranslated crystal (Fig. 6.3). As glide proceeds, the line moves according to the imposed stress, and will retreat if the stress is reversed. The unit displacement of a dislocation is called its Burgers vector (**b**), which usually is the shortest translation distance of the unit cell in the slip plane. The progress of a dislocation restores the atomic structure as it passes, so that changes in grain shape can be achieved without destroying the crystal structure.

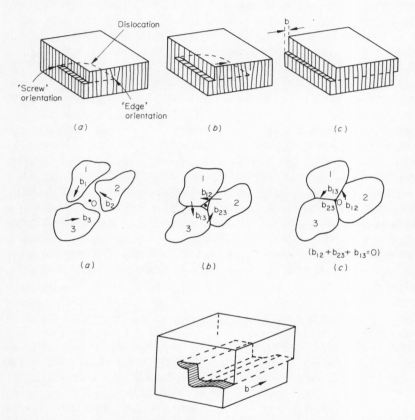

Fig. 6.3 *Top*: spread of translation gliding across a crystal by means of a general dislocation; *middle*: joining of three dislocations to form a node, which is possible when the sum of the three Burgers vectors is zero; *bottom*: diagram showing how a screw dislocation may glide on any plane. After Gilman, pp. 24, 25, 26.[5]

Dislocation lines cannot end within a crystal, because they are the boundaries of surface areas ('slipped areas') and so must either close on themselves to form a 'loop', or end against another surface (such as a grain boundary or another slipped area). Three or more dislocation lines can meet at a point ('*node*') if the sum of their Burgers vectors is zero (Fig. 6.3). Consider three dislocated loops expanding to join and surround a point O (note that each loop lies on the same glide plane, but has a different **b** from the other two). Along the joined areas, **ɔ** becomes the vector sum of the **b**s of adjacent loops, e.g. $\mathbf{b}_{12}=\mathbf{b}_1+\mathbf{b}_2$. The displacement at O can remain zero only if $\mathbf{b}_{12}+\mathbf{b}_{13}+\mathbf{b}_{23}=0$, in which case a *node* can form; otherwise the loops will stay surrounding the point O (as in the middle of Fig. 6.3).

Note that the tangent to the dislocation line is its direction at that point. Usually this is between 0 and 90° of the Burgers vector. The two extremes are called *edge* orientation (edge dislocation), where the dislocation line is normal to its Burgers vector; and *screw* orientation (screw dislocation) where the dislocation line is parallel to its **b** (Figs 6.3, 6.4). Dislocations with intermediate orientations are called *general* (mixed) dislocations (Fig. 6.3).

Screw dislocations can glide on any plane (keeping **b** constant), the process being called *cross-slip* or *cross-glide* (Fig. 6.3). But edge dislocations can glide only on one plane, to maintain a constant length of the 'extra half-plane' (extra lattice plane above or below the dislocation; Fig. 6.4). If the extra half-plane could have an extra atom added on to it or subtracted from it (e.g. by vacancy diffusion) it could *climb* out of its glide plane and be able to glide on another plane. Climb occurs at temperatures high enough to promote diffusion (i.e. it is T-dependent), and at slow strain rates (where more time is available for diffusion). It explains why crystals can be deformed more easily (i.e. are weaker) at higher temperatures and slower strain rates. Climb of edge dislocations causes a shortening or extension of the crystal, providing an additional deformation mode (other than slip and twinning). At slow strain rates (such as those likely to apply to rocks), climb of edge dislocations and cross-slip of screw dislocations are probable mechanisms.

Experimental Detection of Dislocations

Dislocations can be detected in various ways, among them being:

(i) *etch pitting*, in which the distorted core of the dislocation (which is too small to see with the microscope) is preferentially dissolved by suitable chemical reagents, producing a pit that is much bigger than the core, thus revealing the outcrop of the dislocation on the surface being examined;

(ii) *decoration* (precipitation), in which thin threads of impurity atoms can be precipitated along dislocation lines, to make their traces visible in the optical microscope in intensely transverse light ('ultramicroscopy');

(iii) *transmission electron microscopy*, which involves the thinning of the material (including silicates nowadays, by ion bombardment) until electrons can be transmitted. Elastic strains around dislocations cause diffraction of the electrons, so that the dislocations show up as dark or

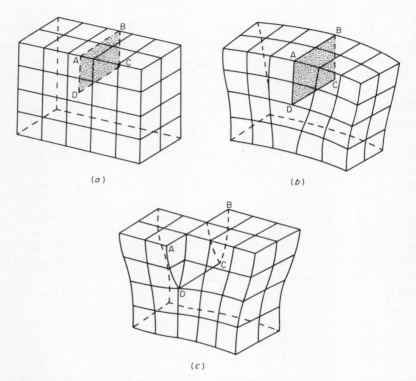

Fig. 6.4   (*a*) Simple undeformed cubic lattice; (*b*) positive *edge* dislocation (DC) with an extra half plane (ABCD) above the dislocation line; (*c*) a left-handed *screw* dislocation (DC) formed by displacing the ABCD faces relative to each other in the AB direction. After Hull (1965). *Introduction to Dislocations*, New York: Pergamon Press.

light lines. This is a very useful approach to experimentally and naturally deformed minerals and rocks, because it gives us direct evidence of the dislocation distribution, and so may tell us something about deformation and recovery mechanisms (Chapter 7).

Mobility of Dislocations

If the stress fields around two nearby parallel dislocations cancel, the dislocations attract each other and so may be mutually annihilated. This can happen if the dislocations are of opposite 'sign' (Fig. 6.4). On the other hand, if the stress fields reinforce, the dislocations repel each other.

However, when a pair of non-parallel dislocations intersect, a *jog* or *step* in the dislocation line is formed in either or both dislocations. This is true of two edge dislocations, two screw dislocations, or an edge plus a screw dislocation. Generally, the length of a jog equals the component of the Burgers vector normal to the other dislocation line. The Burgers vector of the jogged section is always the same as that of the dislocation line of which it is a part. Jogs make dislocations less mobile, because either more work or diffusion (to cause climb) is necessary to move them. The increased jogging of dislocations during plastic deformation (producing dislocation 'tangles' or 'pile-ups') makes an ever-increasing stress necessary to cause further dislocation movement, which is an important cause of work- (strain-) hardening of crystalline grains undergoing deformation at relatively low temperatures and/or relatively fast strain rates.

Production of Dislocations

A very low stress is needed to move a single, isolated dislocation, but the amount of strain produced is also very small (i.e. $= \mathbf{b}$). Macroscopic strain requires the co-operative movement of many dislocations. Severe plastic deformation commonly changes the dislocation density in polygrain material from 10^7–10^8 to 10^{11}–10^{12} lines/cm^2. Some mechanisms must enable this to happen, because we have seen that dislocations are not there for thermodynamic reasons.

Several mechanisms are possible:

(i) *Formation by growth mistakes*, which are common in all crystalline materials. Dislocations formed in this way may move under stress and so assist plastic flow. But many of them are immobilised (*a*) by impurity particles precipitated along them before deformation, (*b*) because they have incorrect Burgers vectors, or (*c*) because they lie in the wrong glide planes for the stress system concerned. So, 'grown-in' dislocations generally are probably inadequate to account for most deformations.

(ii) *Stress nucleation*, which involves the initiation of dislocations at sites of high stress. The nucleation can be either homogeneous (random) in a volume of perfect crystal structure (in which case a very large stress fluctuation is needed) or heterogeneous (which is much more common, being assisted by existing stress concentrations near cracks, precipitates and other defects).

(iii) *Regenerative multiplication*, by which existing dislocations give rise to others in a continuous process during deformation. The two main regenerative mechanisms are as follows:

(*a*) *Frank-Read source*: consider a segment (AB) of dislocation pinned (i.e. immobilised) at both ends by nodes or jogs (Fig. 6.5(*a*)). Because AB

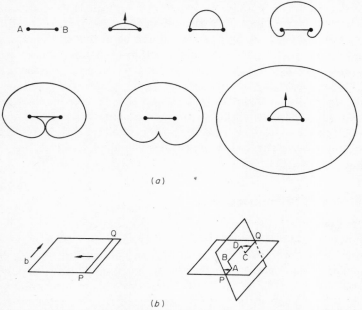

(*a*)

(*b*)

Fig. 6.5 (*a*) Multiplication of dislocations by means of a Frank-Read source involving a pinned dislocation AB; (*b*) Cross-glide of part of a screw dislocation to produce edge dislocations AB and CD that cannot move in the direction of motion of the original dislocation, so that the pinned segment BD can operate as a Frank-Read source. After Gilman, pp. 33, 34.[5]

lies in the glide plane and has the right **b** for glide in the stress field, it can move. When stress is applied, AB bows out and eventually forms a double loop on the other side of the original segment. At the point where the loops meet the dislocations annihilate each other, producing a single mobile loop completely surrounding AB. Now, AB is again free to move, so that the whole process can be repeated, resulting in a succession of loops moving further and further outwards from the original segment.

(*b*) *Multiple Cross-glide*: this is similar to a Frank-Read source, but involves a change of glide plane, so that it is applicable only to initial screw dislocations. Consider a screw dislocation (QP) moving from right to left in a glide plane (Fig. 6.5(*b*)). If a segment of it (BD) cross-glides on to another plane, it forms two segments (AB, CD) that cannot move in the direction of QP (because they are edge dislocations and so cannot move normal to **b**). So, they anchor BD, which then operates as a Frank-Read source. Of course, loops formed in this way can themselves cross-slip and start forming new Frank-Read sources. So, deforming grains have plenty of opportunities for forming new dislocations.

Dislocations in Relation to Work-Hardening and -Softening

Deformation involves a competition between hardening processes, namely the jogging and eventual tangling of dislocations, and softening processes, such as climb, cross-slip, dislocation annihilation, running out of dislocations into voids or grain boundaries, subgrain formation or recrystallisation. The degree to which either process predominates depends on such factors as the temperature, strain rate and availability of 'volatile' components (especially hydroxyl, as far as silicates are concerned, since it tends to break strong O-Si bonds and allow dislocations to move, as discussed towards the end of this chapter and in Chapter 7).

Partial Dislocations[10, 17]

So far I have mentioned only complete (perfect, total) dislocations, which involve the transfer of an atom to a point normally occupied by another atom, so that a perfect crystal structure occurs on each side of the dislocation. But, in many structures, several possible atom sites could provide stable (or metastable) resting places for an atom in a migrating dislocation, without restoring a perfect crystal. If this happens, the dislocation is called a partial dislocation, and it brings about an error in the stacking sequence of atomic structure planes. This is called a *stacking fault*, and is bounded by two partial dislocations that separate it from the rest of the perfectly stacked crystal. The net displacement vector of the two partial dislocations, taken together, is the same as that of a theoretical perfect dislocation, the dissociation into partials being permitted because their Burgers vectors are shorter than that of a complete dislocation. So, the net displacement can take place in two energetically easy steps, rather than one hard one. Of course, this can only happen if the extra energy produced by the resulting stacking fault is less than the energy saved by the formation of the partial dislocations. The production of partial dislocations affects the slip processes operating in a material.

Dislocations in Non-Metals

Factors complicating the structure of dislocations in non-metals are: (i) decreased symmetry of the crystal structure; (ii) increased complexity of the crystal structure; and (iii) bonding characters (charge, directivity, covalency, etc.). The bonding generally is stronger, the unit cells are bigger and more complex, and dislocations are harder to visualise.[14] Edge dislocations may have extra half-'planes' consisting of more than one plane (even of unit cell dimensions); e.g. in NaCl the extra half-'plane' is a double layer of Na^+ and Cl^- ions, needed to maintain electrical neutrality.[10, 14]

The Burgers vectors of total dislocations may be so large that dissociation into partial dislocations may be more favourable, giving rise to numerous stacking faults (extended dislocations) which turn out to be quite common in silicates (see later). The directional nature of covalent or partial covalent bonding (such as in oxides and silicates) makes dislocation movement difficult, and probably accounts for the typically high yield strength of these materials, compared with metals.

Surface Imperfections[1,3,4,10,17]

A surface imperfection (planar defect) involves a change in the order or orientation of atom planes across a boundary. This change may be one of orientation or of stacking sequence of planes. Some planar defects separate crystal orientations so large that they are said to be different grains, the boundaries between them being called grain boundaries (*inter*granular boundaries). All other planar defects are *intra*granular boundaries. The main types of planar defects are as follows:

(i) *Antiphase boundaries.* Crystalline solid solutions tend to have random (disordered) distributions of atoms at higher temperatures. Slow cooling causes certain atoms to select lower-energy sites in the less open crystal structure. For example, this ordering is well shown by volcanic pigeonite (Fig. 6.6), in which lattice sites that are equivalent at high temperatures become non-equivalent at lower temperatures. Because the ordering can start at either of these formerly equivalent sites, ordering on two different 'registers' (each with a 50 per cent chance of occurrence) can take place.[4] The boundary along which domains of the two different registers meet is called an *antiphase boundary* (Fig. 6.6), since the same phase occurs on either side, the boundary merely marking a difference in registry. Antiphase boundaries have been observed in silicates with the transmission electron microscope.

(ii) *Low-angle boundaries* (subgrain boundaries), which separate parts of a grain with only small angular misorientations (up to about 7°). They are made up of dislocation arrays of variable complexity, and generally are formed by recovery either during or after deformation (Fig. 6.7).

Because of the small misorientation across them, their structure is not complex enough to cause much light scattering or preferential deepening during grinding, so that they show up optically as thin to indistinct lines separating areas of slightly different extinction orientations. They generally are relatively immobile boundaries.

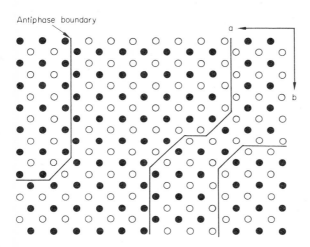

Fig. 6.6 Schematic diagram showing antiphase domains in pigeonite. At higher temperatures both open and filled circles are equivalent lattice sites, but not at lower temperatures. Antiphase boundaries occur between domains ordered on different registers. After Champness & Lorimer.[4]

(iii) *Twin boundaries,* which, in contrast, are *high*-angle boundaries (as are grain boundaries), but which separate two orientations that have a specific symmetrical relationship to each other. The orientations are related by a symmetry element not possessed by the untwinned structure. For example, in albite-law twinning in plagioclase, one half of the twin is related to the other by a 180° rotation about the normal to (010), so that (010) becomes a reflection plane in the twinned structure, although it is not in the untwinned structure. Twin boundaries are not favoured thermodynamically, but are stable over a fairly wide range of conditions, because the atoms in the boundary (twin interface, composition plane) are shared by the structures on both sides of the twin. So, most twin boundaries are of relatively low energy, and, consequently, do not migrate normal to their lateral extent, as grain boundaries do (Chapter 5). They may be formed by growth accidents or by deformation (as mentioned above), but some types of twin that can form by growth cannot form by deformation, because the atom movements involved would be too complex to occur in the relatively short time periods that deformation twins usually take to form.

(iv) *Stacking faults* are boundaries, the crystal structures on either side of which have the same orientation, but one side is translated with

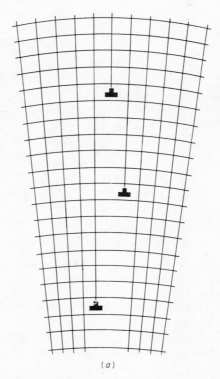

(a)

Fig. 6.7 (a)   Diagrammatic representation of a tilt-boundary (a simple variety of low-angle boundary), showing an evenly spaced array of positive edge dislocations.

respect to the other by a fraction of a lattice vector. As mentioned previously, this faulted surface is separated from the unfaulted structure by a partial dislocation, if the stacking fault ends within a grain. Stacking faults may be produced as growth accidents or by the separation of two partial dislocations during deformation. They are common in silicate minerals (e.g. talc, plagioclase and kyanite). Because of the lack of misorientation across them, they are not visible optically. The area of the stacking fault depends on its energy, and there is a balance between the tendency of the crystal to reduce the area of the faulted region and the tendency of the two partial dislocations to repel one another. A stacking fault will not form if the energy increase would exceed the energy drop due to dissociation of a perfect dislocation into two partials.

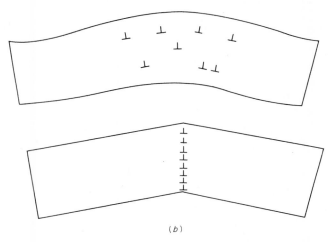

(b)

Fig. 6.7 (*b*) Diagrammatic idea of the formation of a low-angle boundary from a bent grain with random edge dislocations (top), to a grain with a symmetrical tilt boundary (a variety of subgrain boundary), by the glide *and* climb of the dislocations (bottom). After Hull (1965). *Introduction to Dislocations*. New York: Pergamon Press.

(v) *Deformation bands* and *deformation lamellae* are planar defects formed during deformation, and are discussed in Chapter 7, in relation to the deformation of quartz, olivine and plagioclase. *Kink bands* (described above) are sharp deformation bands formed by simple bending. Other deformation bands are more complicated. Deformation lamellae are narrow zones with a slightly different refractive index from the adjacent grain. In some experiments, deformation lamellae have been shown to form parallel to slip planes, and have been used as indicators of slip systems (see later). Other deformation lamellae have complicated structures and may not reflect slip alone.

(vi) *Grain boundaries* are the most general planar defect. A knowledge of their structure is essential to an understanding of recovery and recrystallisation processes. Around 1952 people still argued about whether grain boundaries were amorphous layers or transition lattices, but all recent work favours the second alternative. Most studies of grain boundary structure concern metals, and, as discussed in Chapter 2, the nature of grain boundaries in silicates is very problematical. Any boundary in a pure metal can be thought of as an array òf atoms common to two grains that meet at a relatively high angle (say, greater than about 10°). Statistically, they can be regarded as being of uniformly high energy, compared with low-angle boundaries and twin interfaces. However, certain high-angle boundaries in metals have experimentally detectable lower energies than random grain boundaries, as discussed below.

At certain angular relationships between the grains (even at high angles) certain atom sites coincide at the boundary, and, furthermore, if one of the two lattices is extended to overlap the other, a three-dimensional array (or 'superlattice') of coincidences (*coincidence lattice*) is produced. However, the only atoms common to both lattices are at the boundary itself. These *coincidence boundaries* appear, from certain experiments, to move faster than random high-angle boundaries, although this is intuitively surprising, in view of the known immobility of twin boundaries (mentioned above), which are also examples of high-angle coincidence boundaries (perfect coincidence boundaries, in fact).

The slightest departure from the geometrically exact relationship must completely destroy the coincidence lattices, but boundary coincidence still can be maintained for about 10° away, by distortion (which increases as the misorientation increases). In fact, any boundary may be represented as a special case of the transition between ideal coincidence angles, and the boundary need not be symmetrically arranged between the two lattices.[1]

Experimental evidence shows us why coincidence or near-coincidence boundaries are so mobile. Even high-angle boundaries have 'pipe-like' regions of relatively open structure that permit easier diffusion of impurities in selected directions. This is shown by measured anisotropies of grain boundary diffusion that persist up to misorientations of 45°. Experiments[1] on the effect of impurities on grain boundary mobility show that coincidence boundaries move up to 100 times faster than random boundaries, because they are more ordered and so permit either the removal of impurities along the 'pipe-like' holes or the easy retention of impurities in dislocation cores (so that the impurities can migrate with the boundary). On the other hand, the more widespread, complicated dislocation array of a random boundary would restrict diffusion of impurities and would attract more impurities, which would impose a drag on the boundary. Since impurities are nearly always present (and we could expect them especially in rocks), this could be a major reason for the differential mobility of grain boundaries. Another point of evidence for coincidence boundaries is that the nucleation and growth of twins during grain growth can only occur if new, lower energy (hence coincidence) high-angle boundaries are generated along with the new twin interface (Fig. 6.8).

The relevance of all this to recrystallisation (see below) is that, if boundaries in selected orientations move fastest, then a special crystallographic relationship between the deformed host and the new, recrystallising grains could result, which should have an effect on the preferred orientation of the resulting aggregate.

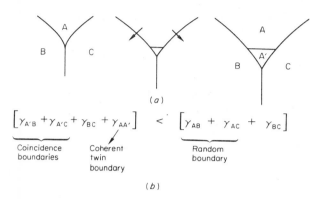

$$\left[\gamma_{A'B} + \gamma_{A'C} + \gamma_{BC} + \gamma_{AA'}\right] \quad < \quad \left[\gamma_{AB} + \gamma_{AC} + \gamma_{BC}\right]$$

Coincidence boundaries   Coherent twin boundary   Random boundary

(*b*)

Fig. 6.8(*a*) Diagrammatic representation of the nucleation and growth of a twin during ordinary grain growth in the solid state. Grain A is growing at the expense of grains B and C. Note that the newly formed twin boundary (AA') cannot move normal to itself (because it is of very low energy, and hence is stable), but lengthens parallel to itself during continued growth of grain A. The energy requirements for the twin to be stable are shown in (*b*).

Effect of Grain Boundaries on Deformation (Intergranular Deformation)

At lower temperatures, grain boundaries provide obstacles to the movement of dislocations, and so increase the rate of strain-hardening. So, single crystals generally deform more readily than polygrain aggregates of the same phase. Because they are barriers to dislocation motion, grain boundaries are locations of high stress concentration, and if the grains cannot accommodate to the strains of their neighbours (i.e. if they have insufficient slip systems under the prevailing conditions), voids or cracks occur at the boundaries. This could lead to cataclasis, as discussed previously. Therefore, we might expect to see evidence of voids or cracks in polygrain silicates deformed at low temperatures and/or rapid strain rates, especially those of lower symmetry (e.g. layer silicates).

At higher temperatures and slower strain rates, grain boundaries (especially in ceramic materials, and so probably also in many minerals) become weak, so that grain boundary sliding takes place.[4] This generally produces voids along the boundary, unless recovery processes cause corrugations that inhibit the sliding, or unless 'pressure solution' (see later) can fill voids as soon as they begin to open. Growth and coalescence of the voids leads to intergrain fracture. So, in natural rock deformation under metamorphic conditions we might expect to see evidence of voids formed by grain boundary sliding, especially in lower-symmetry minerals, but probably also in others. Transmission electron microscopic examination of some deformed quartz, plagioclase and biotite indicates that such voids are present (Stan White, personal communication).

Voids formed by grain boundary deformation are a potentially important means of removing the volatile products of prograde devaporisation reactions, and of introducing volatile reactants in hydration reactions (Chapter 2). The common localisation of retrograde reactions in deformation zones could be due as much to this mechanism as to the internal deformation of metastable grains of the high-grade assemblage (Chapter 3).

Recovery and Recrystallisation[8,9,11,15]

Recovery is the set of processes that decreases work **(strain) hard**ening, without the development of high-angle boundaries. No nuclei of new grains are produced. Instead, recovery generally forms strain-free, relatively sharply bounded *subgrains*, the process being called 'polygonisation'. Note that this word should not be used for the production of polygonal grains (Chapter 5). The subgrains are separated by low-angle boundaries (subgrain boundaries). Dislocations migrate and climb into stable arrays (walls) arranged transverse to their active glide planes (Fig. 6.7). Also, a general reduction in dislocation density takes place by mutual annihilation of moving dislocations, and by running out of dislocations into grain boundaries and voids. Recovery is slow at low temperatures, and increases in both rate and amount with increasing temperature. It is so rapid above a critical temperature that no strain-hardening occurs; i.e. dislocations climb and cross-slip so readily that pile-ups fail to develop. *Cold-working* is deformation at temperatures low enough to cause an appreciable amount of strain-hardening. *Hot-working* is deformation at temperatures too high for strain-hardening, e.g. above 500°C for calcite.

Recrystallisation occurs at temperatures greater than those at which recovery occurs (i.e. greater than about half the absolute melting temperature), and releases any remaining strain energy. Strictly, recrystallisation is simply the nucleation and migration of high-angle boundaries, and so could apply to situations with or without phase changes. Boundaries move in response to stored strain energy, so that strain-free grains replace strained material. The result is an aggregate of strain-free grains, and, if further growth and boundary adjustment occur, the process is called *grain growth,* as described previously (Chapter 5). *Annealing recrystallisation* ('primary recrystallisation') is the recrystallisation of a formerly cold-worked crystalline material. *Syntectonic recrystallisation* is recrystallisation during deformation (e.g. in hot-working). Though these terms can be used in experiments for which the controlling conditions are known, we must be very careful in applying them to rocks; generally we are trying to find out whether these processes took place.

The recrystallisation process involves two steps: (*a*) nucleation (the formation of stable strain-free areas with high-angle boundaries suitable for rapid growth), and (*b*) growth (the expansion and eventual impingement of the stably nucleated grains). Three main mechanisms of nucleation have been proposed for metals:

(i) *Random (homogeneous) nucleation* (Chapter 3), which generally is discounted, because the minimum size of the stable nucleus (about 1μ) and the activation energy for nucleation are both too large to be related to random thermal fluctuations. So, the nuclei appear to be present in the deformed grain; i.e. they are produced by the deformation itself.

(ii) *Preformed or subgrain nucleation,* in which subgrains form in areas of high strain (especially along deformed grain boundaries and kink band boundaries), and selected ones grow by absorbing their neighbours, increasing their misorientation with respect to the original host-grain as they go, by accumulation of collected dislocations into their boundaries. Eventually these boundaries attain a high enough misorientation with respect to neighbouring material for them to move rapidly. At this stage the subgrains have become stable nuclei suitable for recrystallisation.

(iii) *Bulge nucleation,* which needs only a high-angle boundary (e.g. a kink band boundary) formed during deformation; i.e. the formation of sub-grains is not necessary. If there is a difference in dislocation density on either side of a high-angle boundary, heating causes the boundary to move towards the more deformed side, absorbing dislocations as it goes. Local fluctuations in dislocation density cause bulges in the boundary and these can act as recrystallisation nuclei. This type of nucleation is common in moderately deformed metals annealed at temperatures below those required for dislocation climb and subgrain formation. Bulging of kink boundaries also appears to occur in quartz.[8]

All these mechanisms are driven only by release of stored plastic strain energy. If, as commonly happens in minerals, recrystallisation is accompanied by chemical change, the chemical free energy can contribute to the driving force for recrystallisation, as discussed in Chapter 7.

After nucleation (i.e. the development of a mobile high-angle boundary), growth of the strain-free areas (new grains) occurs by movement of the boundary, which collects dislocations remaining after recovery as it goes. As noted above, coincidence or near-coincidence boundaries move faster than random boundaries, so that grains of specific orientations may result. Eventual impingement of the high-angle boundaries produces an aggregate of strain-free grains with irregular to smooth, random boundaries. Although strain energy has been minimised at this stage, interfacial free energy can be still high, so that further grain growth can occur if the temperature stays high enough (Chapter 5).

Development of Crystallographic Preferred Orientation by Recrystallisation

Since recrystallisation can occur either after deformation or during deformation (syntectonic recrystallisation), the possibility should be considered that preferred orientations could develop by the recrystallisation process itself. Possible controlling mechanisms[6] include:

(i) *differential strain energy*, in which deformed grains with less stored strain energy grow preferentially, eliminating grains in other orientations favourable for greater deformation;

(ii) *oriented nucleation*, in which nuclei form in restricted orientations, possibly controlled by stress or strain during deformation;

(iii) *oriented growth*, in which nuclei of all orientations form, but the higher mobility of coincidence boundaries causes preferential growth of grains in some orientations relative to the deformed host; alternatively, the selection of suitable growth orientations is controlled by the strain or the operating stress during syntectonic recrystallisation.

Development of Preferred Orientations by Deformation

Strong plates and rods in a flowing homogeneous medium[15] tend to rotate so that their long axes become parallel to the axis of maximum extension. One long-standing interpretation of the preferred orientation of mica in slates and schists involves such 'mechanical rotation', with or without intragranular deformation. Some experimental work on the deformation of layer silicate aggregates supports this idea (Chapter 7), and it would be expected to apply especially where there is a penetrative pore fluid at relatively high pressure,[15] thereby forcing grains apart. However, in higher-grade rocks (e.g. mica schists), resistance to rotation would probably be considerable, so that intragranular deformation may be more important.

In cold-worked crystalline materials, preferred orientations can be produced by intragranular deformation, accompanied by rotation of each grain in space in a sense opposite to that of intragranular gliding. Steady-state orientations of grains tend to be produced, which stay approximately the same as strain increases. The symmetry of the preferred orientation is related to that of the deformation process.[15] Specific examples of this kind of preferred orientation will be discussed in Chapter 7, especially with regard to quartz.

Some Generalities

The mechanical behaviour of ductile crystalline materials varies greatly with the conditions controlling the various internal processes. At low temperatures and relatively fast strain rates, intracrystalline slip predominates, and work-hardening occurs as dislocations become jogged and tangled. Optically, the grains generally appear to have been flattened, and show such features as deformation lamellae, deformation bands and twinning. Dimensional and crystallographic preferred orientations are due mainly to intragranular slip and accompanying grain rotations.

At higher temperatures and slower strain rates cross-slip and climb of dislocations enable the rate of recovery to keep pace with the rate of work-hardening, so that the material is weaker. Optically, deformation lamellae and deformation twins are less common, being replaced by such features as subgrains, deformation bands, serrated (migrated) boundaries (both grain boundaries and kink boundaries) and recrystallisation. All these are reflections of increased atom mobility. The transmission electron microscope shows general reductions in density of dislocations, accompanied by strong tendencies for dislocations to be arranged in subgrain boundaries.

Depending on the nature of the minerals and the PTX conditions, rocks undergo varying degrees of the above deformational processes. Water (and probably other 'volatile' components) has a strong effect on the mobility of dislocations, and so affects recrystallisation and mechanical properties. It appears to be common in minerals generally regarded as anhydrous (e.g. kyanite, andalusite, sillimanite, garnet, olivine, feldspar and diopside).[16] The water appears to be present as hydroxyl ions occupying oxygen sites associated with vacancies or charge imbalance in cation sites.[16] Though the amounts of (OH) ions are very small, they are probably effective nevertheless, because of the experimentally observed, pronounced effects on strength and recrystallisation ability of quartz and olivine produced by only minute amounts of water in the minerals' structure (Chapter 7).

Also, many metamorphic temperatures are high enough and strain rates probably are small enough to cause relatively high recovery rates during deformation, so that many, if not most, metamorphic rocks should show at least some evidence of recovery and recrystallisation, which should not automatically be ascribed to post-deformational heating,[9] as has been done often in the past.

Appendix: Methods of Detecting Slip Systems

Slip Bands   Raleigh[12] has devised an ingenious method for observing slip bands, which are invisible in thin sections of transparent materials. The cores used in deformation experiments were split longitudinally; both faces were polished and thin platinum foil was placed between them. After a few per cent strain, the polished surfaces showed slip markings (bands) or twinning, owing to the extreme weakness of the platinum, and were photographed in reflected light. Then the surfaces were lightly ground, cemented to glass slides, and made into thin sections, which enabled optical vectors, cleavages and twin planes to be oriented with a universal microscope stage. In this way, the crystallographic indices of the slip bands (active planes of slip) can be inferred. The slip direction can be determined from other deformational heterogeneities such as kink bands.

Kink Bands (Fig. 6.2)   The slip plane contains both the axis of external rotation (which is parallel to the intersection of the slip plane and the kink band boundary), and the slip direction (which lies normal to the intersection of the kink band boundary and the slip plane). The axis of external rotation is a direction common to both rotated and initial lattice orientations. Where external rotations are small (less than about 10°), determination of the axis of external rotation is made less accurate because the rotation angle is only a few degrees larger than the errors involved in measurement of optical vectors.

Deformation Lamellae   These form at angles of high resolved shear stress in experiments, and have been shown by electron microscopy to be due to collections of dislocations in slip planes, so that measurement of their orientation gives the slip plane, though generally not the slip direction, unless they are associated with kink bands (see above), or unless electron microscopy is used to indicate the Burgers vectors of dislocations involved. In addition, if several co-zonal deformation lamellae are present, the zone axis is the slip direction. However, natural deformation lamellae in quartz do not necessarily lie in slip planes, so that inferences of slip systems from natural deformation lamellae should not be undertaken without transmission electron microscopy (Chapter 7).

Decoration of Dislocations This has been used to detect slip systems in high-temperature experiments in olivine, where deformation lamellae are absent. Carter and Ave'Lallemant successfully tried the ingenious technique of decorating the dislocations during the deformation experiments.[2] Because dislocations (especially edge dislocations) can reduce their core energy by attracting certain foreign atoms, pre-heating and deforming the olivine samples with various metallic powders (especially manganese and magnesium) causes minute metal particles to be precipitated at dislocation sites, thus revealing the active slip planes.

Direct Observations of Dislocations Direct observation with the electron microscope provides an excellent means of inferring the slip systems, namely by determination of the Burgers vectors of the dislocations.

References

1 Aust, K. T. and Chalmers, B. (1970). Structure of grain boundaries. *Metallurgical Transactions*, **1**, 1095–104.
2 Carter, N. L. and Ave'Lallemant (1970). High temperature flow of dunite and peridotite. *Bull. Geol. Soc. America*, **81**, 2181–202.
3 Chalmers, B. (1959). *Physical Metallurgy.* New York: J. Wiley & Sons, Inc.
4 Champness, P. E. and Lorimer, G. W. (1971). An electron microscopic study of a lunar pyroxene. *Contribs. Mineralogy & Petrology*, **33**, 171–83.
5 Gilman, J. J. (1961). Nature of dislocations, in *Mechanical Behavior of Materials at Elevated Temperature*, ed. Dorn, J. E. 17–44. New York: McGraw-Hill Book Co.
6 Green, H. W., Griggs, D. T. and Christie, J. M. (1970). Syntectonic and annealing recrystallisation of fine-grained quartz aggregates, in *Experimental and Natural Rock Deformation*, ed. Paulitsch, P. 272–335. Berlin-Heidelberg-New York: Springer-Verlag.
7 Hayden, H. W., Moffatt, W. G. and Wulff, J. (1965). *The Structure and Properties of Materials.* Vol. III. *Mechanical Behavior.* New York: J. Wiley & Sons, Inc.
8 Hobbs, B. E. (1968). Recrystallization of single crystals of quartz. *Tectonophysics*, **6**, 353–401.
9 Hobbs, B. E., Means, W. D. and Williams, P. F. (1975). *An Outline of Structural Geology.* New York: J. Wiley & Sons, Inc.

10 Kingery, W. D. (1960). *Introduction to Ceramics*. New York: J. Wiley & Sons, Inc.
11 Paterson, M. S. (1969). The ductility of rocks, in *Physics of Strength and Plasticity*. Mass. Inst. of Technology.
12 Raleigh, C. B. (1968). Mechanisms of plastic deformation of olivine. *J. Geophys. Research*, **73**, 5391–406.
13 Read, W. T. (1953). *Dislocations in Crystals*. New York: McGraw-Hill Book Co.
14 Spry, A. (1969). *Metamorphic Textures*. Oxford: Pergamon Press.
15 Turner, F. J. and Weiss, L. E. (1963). *Structural Analysis of Metamorphic Tectonites*. New York: McGraw-Hill Book Co.
16 Wilkins, R. W. T. and Sabine, W. (1973). Water content of some nominally anhydrous silicates. *Amer. Mineralogist*, **58**, 508–16.
17 Moffatt, W. G., Pearsall, G. W. and Wulff, J. (1964). *The Structure and Properties of Materials*. Vol. I. *Structure*. New York: J. Wiley & Sons, Inc.

Chapter 7

Deformation, Recovery and Recrystallisation of some Common Silicates

Introduction

To illustrate the general processes discussed in Chapter 6, we will now look at experimental work on some of the common silicate minerals, and try to relate this to observations on natural metamorphic rocks. The minerals chosen as examples are quartz, olivine, layer silicates and feldspar (especially plagioclase). The aim is to describe and try to interpret (i) microstructural features and (ii) preferred orientations, produced by deformation, recovery and recrystallisation. This should give us an indication of the extent to which we can relate microfabric features to metamorphic conditions, and may also tell us something about the processes controlling the deformation and recrystallisation of metamorphic minerals. I should emphasise that this kind of work is in its early stages, and no doubt many of the ideas expressed here will need alteration in the near future.

Before experimental results were available, structural petrologists based their interpretations on speculative mechanisms (e.g. hypothetical slip systems in quartz). Experimental deformation, coupled with careful optical and transmission electron microscopic examination of the products, has not only revealed processes, but is guiding us in the description and interpretation of natural deformation features.

QUARTZ

Introduction

Evidence of deformation and recovery in natural quartz is common, in the form of undulose extinction, deformation bands and deformation lamellae. Evidence of recrystallisation of deformed grains to initially finer-grained aggregates of optically strain-free grains is also common in natural quartz. Experimental results give us some information on the conditions of this deformation and recrystallisation, but considerable care must be exercised in interpreting the results. For example, most experiments involve axial compression, whereas natural deformations may be more complicated, and strain rates are much larger in experiments than in natural rock deformation. Furthermore, because the direction of shortening is known in these experiments, some idea of the stress field may be inferred, whereas generally this is impossible in the investigation of naturally deformed aggregates. Another problem is that all quartz in rocks exposed at the earth's surface is α-quartz,[21] although it may have been the β polymorph during deformation and have undergone deformation and recrystallisation processes different from those applying to α-quartz. Because the $\alpha \rightleftharpoons \beta$ transition is displacive (Chapter 2), the β-form cannot be quenched-in by cooling, so that optical or X-ray evidence is inadequate to indicate its former existence in metamorphic rocks.

Most experiments have been carried out at a relatively high confining pressure, which is necessary to avoid brittle deformation. Early experiments (before about 1963) involved deformation of dry quartz, which remained strong and generally brittle. Introduction of water, however, has made ductile deformation possible at experimentally reasonable strain rates. In fact, experiments on quartz with varying (OH)-content show that at a critical temperature, which decreases with increasing (OH)-content, quartz becomes markedly weaker ('water-weakening' effect).[26, 27] The amounts of structurally bound water needed are small, being of the order of 100 ppm of hydrogen. Moreover, the presence of structural water in both single crystals and polygrain quartz lowers the temperatures of recrystallisation, increases the rate of recrystallisation and produces larger grain-sizes. The reason (mentioned briefly in Chapter 6) is probably that recovery and recrystallisation in quartz depend on dislocation climb, involving diffusion of oxygen and silicon atoms away from the dislocation cores. This process is assisted by the hydrolysis of O-Si bonds at a dislocation core, which involves the trapping of diffusing 'water molecules' by the dislocation, thus forcing silicon and oxygen to diffuse away from the site. The diffusing water probably is released, during heating, from hydrolysed silicon-oxygen bonds; the existence of such bonds in water-bearing quartz is suggested by infra-red absorption spectroscopy. This hydroxyl-induced climb of dislocations appears to be a major mechanism of 'hydrolytic weakening' in quartz, in that recovery is responsible for the weakening.[27] It should be increasingly important in the slow deformation of quartz expected in most natural situations, where the strain rates would be much smaller than in experiments. It may account for the production of micro-structures typical of 'hot creep' (i.e. time-dependent deformation at temperatures greater than half the absolute melting temperature) in natural quartz inferred from geological evidence to have been deformed at temperatures below the conventional experimental hot creep limit.[58, 60]

Microstructural Features of Deformed Quartz

Experimentally and naturally deformed quartz shows microscopic features of heterogeneous deformation, namely:

(i) *extinction bands* (undulose extinction),[13] which are bands generally elongate sub-parallel to [0001], formed by bending of the lattice in response to heterogeneous slip on (0001). Arbitrarily, the radius of curvature is taken to be comparable with or larger than the half-width of the reoriented zone.

(ii) *deformation bands* (kink bands),[13, 15, 28] which are similar to extinction bands, but with radii of curvature arbitrarily taken to be small compared with the half-width of the reoriented zone. Optically, they grade into extinction bands, in fact and by definition. They vary considerably in orientation, but most lie sub-parallel to [0001], these being formed by slip on (0001) accompanied by bending or kinking of the slip planes. Slip parallel to [0001] accounts for bands sub-parallel to (0001) and slip on other systems is needed to account for bands in other orientations. Deformation bands terminate either inside or at grain boundaries.

(iii) *deformation lamellae*,[3, 11, 13, 15, 33] which are narrow, sub-parallel planar or lenticular features in various orientations. They terminate inside grain boundaries and commonly are normal to extinction bands. Optically they are visible as regions of slightly different refractive index from that of the adjacent quartz, but their exact nature and origin appear to be variable.[23, 33, 60]

The orientations of deformation lamellae vary with temperature, confining pressure and strain rate (Fig. 7.1), though, in certain ranges of external conditions, lamellae of all orientations occur ('relatively nonselective' field of Fig. 7.1).[3, 11] The main lamellae orientations are: basal (0–5° between (0001) and the lamellae), sub-basal II (6–15°), sub-basal I (16–30°) and prismatic (81–90°). In naturally deformed quartz, sub-basal I lamellae are common, as would be expected under geologically reasonable strain rates of around 10^{-14} sec^{-1} (Fig. 7.1).

However in some deformed quartzites from the central Swedish Caledonides prismatic lamellae are very common (Chris Wilson, personal communication), indicating high deformation temperatures (Fig. 7.1). Basal lamellae occur only in shock- (impact-) metamorphosed rocks (i.e. in deformation involving fast strain rates).

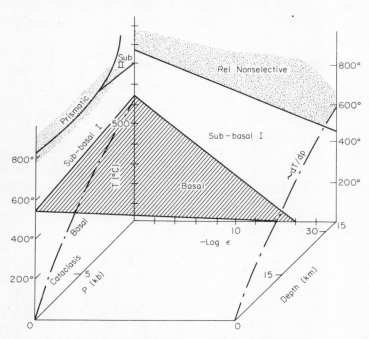

Fig. 7.1 Orientations of deformation lamellae produced experimentally in quartz as functions of temperature, confining pressure and strain rate, extrapolated to geologically reasonable strain rates. After Carter, p. 5520, © by American Geophysical Union.[11]

(iv) *Brazil-law twins*,[21,32] which involve the juxtaposition of regions of left- and right-handed quartz, the rotation of the plane of polarisation produced being so small that the twins are not visible in sections of normal (0·03 mm) thickness, only in very thick slices. Such twinning can be produced by growth or deformation, the second mechanism involving relative displacement of one sub-individual compared to the other, although the crystallographic axes remain parallel. Thin multiple Brazil twins parallel to (0001) of definite deformation origin (as opposed to growth origin) have been produced in quartz single crystals plastically deformed in compression at 500°C and 15 to 20 kb confining pressure, in such a way as to produce a large resolved shear stress along an *a*-axis in (0001). Mechanical Brazil twinning appears to be favoured by relatively low-temperature deformation at rapid strain rates in the absence of structural water, i.e. when the quartz crystals are strong. It does not occur in deformed hydrous quartz and is uncommon in deformed natural quartz. Brazil twins are not removed by heating above the $\alpha \rightleftharpoons \beta$ inversion temperature.

(v) *Dauphiné-law twins,* [21,50,51,52] in which sub-individuals are related to each other by a rotation of 180° about [0001], the hand of the structure being unchanged, so that the twins are invisible optically. The crystal axes remain parallel, but the polarities of the a-axes are reversed, so that positive forms become negative ones, and vice versa. Dauphiné twinning is common in some natural quartz, the composition planes generally being complex, but trending roughly parallel to [0001]. It has been produced by experimental deformation in the α-quartz field (the structure of β-quartz being too symmetrical), but no permanent strain results. The shear stresses needed are very small (of the order of only 20 bars at 400°C), because the twinning is displacive, in that no Si-O bonds are broken. So, it would be expected to occur rapidly during elastic deformation. Because the pattern of Dauphiné twinning reflects the symmetry of the imposed stress system, the preferred orientation of positive and negative crystal forms could indicate the latest elastic stress distribution in an α-quartz aggregate. Dauphiné twinning disappears on heating quartz above the 573° inversion temperature to β-quartz, owing to the higher symmetry of β-quartz prohibiting its existence. On cooling, it may or may not reappear, depending on the elastic strain induced by cooling, and any external stresses operating.

Experimental Deformation, Recovery and Recrystallisation of Quartz

Experimental deformation of coarse-grained quartzite has produced the following general results: [52,53]

(i) *Lower temperatures and faster strain rates.* At small strains, the deformation is very heterogeneous, producing narrow sharp deformation (kink) bands and basal deformation lamellae (implying (0001) slip), without recrystallisation. The edges of grains deform more than the centres. At intermediate strains, the deformation bands are wider, branched and less clearly defined, involving more of each grain in the deformation. Basal slip alone cannot achieve large enough strains, so that slip systems (and associated deformation lamellae) develop in other orientations. Some grains are flattened, but the deformation is so heterogeneous that the original grain boundaries cannot be identified easily. At larger strains, the long axes of elongated grains and possible fragments define a foliation normal to σ_1. Some long, ribbon-like grains are present. Deformation lamellae of many orientations occur, different sets occurring in different parts of the same grain. Some relatively undeformed augen with undulose extinction remain, these being oriented with c either parallel or normal to σ_1, so that they are unsuitable for basal (and, to a less extent, prismatic) slip.

(ii) *Moderate temperatures and slower strain rates.* At temperatures of about 700°C and strain rates of around 10^{-7} sec^{-1}, recovery (dynamic recovery) and recrystallisation (syntectonic recrystallisation) begin to occur, so that work-hardening decreases, and large strains can be obtained with low to moderate differential stresses. The presence of some deformation lamellae (mostly basal), however, indicates that intragranular slip still contributes to the deformation. The recrystallised grains are small and occur along boundaries of original deformed grains. Unrecrystallised parts of grains have sutured boundaries and show smooth undulose extinction instead of sharp deformation bands. With increasing strain the original grains become flatter and more elongate.

(iii) *High temperatures and slower strain rates.* Quartzite deformed at temperatures of 800°C and above, and strain rates of about 10^{-7} sec^{-1}, recrystallises extensively or completely. Deformation lamellae are uncommon and are mainly basal or prismatic. Some augen remain at high strains, but most of the original grains have been transformed into an aggregate of small recrystallised grains and subgrains. Little work-hardening occurs, so that very low differential stresses (around 1 kb) are needed to cause deformation.

Nucleation of new grains initially is (i) intergranular, and subsequently (ii) intragranular (occurring along kink boundaries and other high-strain localities).

(i) *Intergranular recrystallisation.* As in olivine (see later), intergranular recrystallisation takes over, eventually replacing the old grains by polygonal to irregular, optically strain-free grains with orientations related to the experimentally known stress field. Syntectonic recrystalisation of very strongly deformed quartzite (the axial compression exceeding 50 per cent) produces microstructures identical to those in natural quartz-mylonites, namely ribbons of finely recrystallised quartz wrapping around lenticular, flattened augen of residual deformed quartz grains. Many petrologists have interpreted such natural microstructures as being the result of mechanical 'crushing' or 'granulation' of originally coarser-grained quartz (i.e. brittle deformation). However, the experimental microstructures are formed without loss of cohesion, although this can be interpreted as either recrystallisation without fracture, or sintering and recrystallisation of fragments after cataclasis without gross fracturing (i.e. ductile cataclasis; Chapter 6).

Experimental syntectonic recrystallisation of very fine-grained quartz aggregates (flint and novaculite)[24] causes an increase in grain-size and also may produce porphyroblastic microstructures, which may represent secondary (exaggerated) grain growth (Chapter 5). At lower temperatures and higher strain-rates the new grains are elongate normal to the experimental σ_1 and show internal optical deformation features. At higher temperatures and slower strain-rates, the new grains are polygonal, larger, more equant and show much less optical evidence of intragranular strain. Syntectonic recrystallisation of flint in the β-quartz field produces a bimodal size distribution, many of the larger grains being bounded by faces of the form $\{10\bar{1}1\}$. Static annealing in the α-quartz stability field of experimentally deformed flint causes continued growth of grains recrystallised during the earlier deformation without fresh nucleation. Grain shapes are more equant than for syntectonic recrystallisation, and grains are strain-free. Annealing in the β-quartz field gives similar results, except that some grains have square, rather than polygonal, outlines, and these may be bounded by planes of the form $\{10\bar{1}1\}$; they appear to grow faster than grains of other orientations, as discussed later.

(ii) *Intragranular recrystallisation.* This has been investigated in experiments[28] on deformed single crystals of quartz at 300–400°C and 10–15 kb confining pressure. No recrystallisation occurs unless trace amounts of structural (OH) and/or free water are present, although some recovery occurs by means of the untangling and migration of dislocations to form cell walls (subgrain boundaries).[31] The experiments are of three types, namely:

(a) *Annealing* experiments, in which the specimens are deformed at relatively low temperatures and then annealed under hydrostatic stress at high temperature. New quartz grains nucleate along the boundaries of kink bands (deformation bands) and grow faster parallel to the boundaries than normal to them, producing elongate grains. Annealing of experimentally deformed, hydrous, natural quartz at high temperatures produces an aggregate of polygonal grains. Synthetic (hydroxyl-bearing) quartz recrystallises at much lower temperatures, but at a much slower rate, which may be due partly to its relative weakness, so that insufficient strain energy is stored to promote the recrystallisation.[28] Recrystallised areas show only a few dislocation networks (and small bubbles linked by dislocations) in the electron microscope.[31]

(*b*) *Stress annealing (relaxation)* experiments, in which the specimens are loaded to high differential axial compressive stresses at relatively low temperatures, after which the temperature is raised rapidly while the differential load remains. This causes a relaxation (decrease) of differential stress; the amount of relaxation increases with increasing temperature and (OH)-content. Nuclei grow along kink boundaries (which become increasingly misorientated during relaxation), either as bulges in the boundaries or chains of discrete euhedral grains along the boundaries. Continued heating increases the grain-size, the euhedral grains continuing to retain {10$\bar{1}$1} and {10$\bar{1}$0} grain boundaries, implying that faces of these forms migrate relatively slowly. Prismatic faces predominate in large grains.[28] Electron microscopy of these experimentally recrystallised quartz aggregates shows only a few scattered dislocations and rare small bubbles (which always form when hydrogen-bearing quartz is heated above 500°C).[31]

(*c*) *Syntectonic recrystallisation* experiments, in which specimens are deformed to high strains at constant strain rates (of $10^{-5, -6}$ sec^{-1}) and high temperatures (400–900°C). No discrete grains grow from submicroscopic nuclei, but subgrains form in deformation bands at low strains, accompanied by variable migration of the subgrain boundaries. These gradually increase their relative misorientations with increasing strain, until an array of diversely oriented grains with sharp boundaries is produced. Probably bulge nucleation is responsible for the initial boundary migration.[28]

Electron microscopy suggests that, in these constant strain-rate experiments, work-hardening and recovery are competing processes, so that as the deformation temperature is raised the dislocation density decreases, although isolated dense tangles are still present in specimens deformed at 600°C. Generally, with increasing temperature of deformation, the dislocations straighten and interact to form 3-fold nodes. Syntectonically recrystallised quartz shows sharp grain boundaries and very few isolated dislocations in the electron microscope.[31]

Natural Deformation, Recovery and Recrystallisation of Quartz

Naturally deformed quartzites commonly show evidence of heterogeneous intragranular deformation, namely undulose extinction, subgrains, deformation bands and sub-basal deformation lamellae. Conventionally, these optical features have been interpreted as being due to relatively small strains after the general recrystallisation of the rock, which may be true. However, such features are also common in the experimental creep deformation of other crystalline materials, so that an alternative explanation is that they formed during the general deformation and possible syntectonic recrystallisation of the rock.[58, 59, 60]

Transmission electron microscopy of some deformed quartzites[30, 58, 59] has revealed less complexity of lattice defects than expected from the optical heterogeneities mentioned above. The most common defects are (*a*) cellular walls (subgrain boundaries) of discrete arrays of dislocations, and (*b*) individual dislocations (with or without a few tangles) inside the cells. These features, together with the general absence of pinned dislocations, are taken by some people as evidence of creep deformation (probably including syntectonic recrystallisation),[58] and by others as evidence of post-tectonic (annealing) recrystallisation.[30] Subgrains visible optically are revealed by the electron microscope as areas bounded by dislocation walls with misorientations across them of much more than 1°. Smaller subgrains bounded by low-angle walls (1° or less) may occur inside the larger subgrains, causing segmented undulose extinction within the larger subgrains when viewed in the optical microscope.[60] The undulose extinction, kink bands and deformation lamellae are all formed by essentially similar dislocation structures, and, at least in their present state, they appear to have been formed by recovery processes.[60] This means that we should be wary of inferring deformation-recrystallisation histories of naturally deformed quartz on the basis of optical evidence alone. Even quartz grains without any optical evidence of strain may show many dislocations in the transmission electron microscope.

Electron microscopic examination of other natural metaquartzites shows areas of general or localised, dense, tangled dislocations, but no evidence of brittle deformation.[30] These tangles are typical of experimentally strain-hardened quartz and metals. The lack of complete recovery in these specimens is surprising, especially since the deformation is believed to have occurred in the Carboniferous, which should have provided ample time for movement of dislocations into more stable arrangements. Perhaps this is a good example of how cooling below a threshold temperature can cause a thermodynamically favoured reaction to go so slowly as to bring about no significant change, even in a long period of geological time (see Chapter 3).

Sutured (corrugated) grain boundaries are relatively common in natural deformed quartz aggregates, and these could be due to (i) strain-induced grain boundary migration during syntectonic recrystallisation, (ii) grain boundary sliding during creep deformation or (iii) pressure-solution along stylolytic surfaces (Chapter 8).

Also common are various stages in the replacement of deformed quartz by finer-grained, optically strain-free aggregates, which may be due to post-tectonic or syntectonic recrystallisation, on the basis of interpretation of optical microstructure. For example, deformed quartz in a Central Australian mylonite zone[6] shows optical features indicative of recovery and recrystallisation, namely subgrains, new strain-free grains and serrated kink and grain boundaries. Recrystallisation occurs preferentially in more highly strained grains, and occurs either along grain or kink boundaries (associated with lobes or serrations apparently formed by bulge nucleation) or by apparent coalescence of subgrains. Because these subgrains are elongate parallel to the new mylonitic foliation, and because the new (recrystallised) grains also are elongate and show strain shadowing (undulose extinction), the recovery and recrystallisation probably took place during the deformation (dynamic recovery and recrystallisation). In the more strongly deformed rocks of the mylonitic zone, the quartz subgrains are smaller and this may possibly be related to an increased strain rate, arguing by analogy with experiments on metals.

Annealing recrystallisation (under hydrostatic stress conditions) is possible in very fine-grained undeformed quartz (e.g. flint), owing to its relatively large grain-boundary energy. But, strain energy is probably necessary to promote recrystallisation in relatively coarse-grained quartzite, in which the grain-boundary energy is relatively small.

In some natural quartzites, secondary recrystallisation has been inferred by interpretation of microstructures showing scattered large, irregularly shaped, optically strain-free quartz grains in a finer-grained polygonal, equally strain-free aggregate.[61] More advanced stages of this inferred secondary recrystallisation pass eventually into aggregates of interlocking, large quartz grains with no residual small grains, provided mica grains (which could restrict boundary migration: Chapter 5) are absent. These large grains are very irregular in shape in three dimensions, with highly lobate, curved boundaries.

Preferred Orientation

By Deformation   The Von Mises Criterion (Chapter 6) for quartz can be satisfied if climb systems are added to the four independent slip systems produced by a and c slip, or if $<a+c>$ slip occurs on pyramidal planes, remembering that pyramidal slip operates only at higher temperatures or slower strain rates. In experimentally deformed coarse quartzite, crystallographic preferred orientations of original grains have been produced apparently by mechanisms involving intragranular slip (with superimposed mechanical Dauphiné twinning, described previously).[50, 51, 52, 53] At lower temperatures or faster strain rates, the c-axes of original (unrecrystallised) grains tend to lie parallel to σ_1. At higher temperatures or slower strain rates, they form small-circle girdles about σ_1. The opening angles of these girdles increase with increasing temperature and decreasing strain rate, varying from about 15° to 45°. At any temperature and strain rate, the strength of the preferred orientation increases with increasing strain. For example, residual augen in experimentally produced 'quartz-mylonites' show a strong tendency for their c-axes to lie parallel to σ_1. At equivalent strains, the c-axis maxima are weaker than the small-circle girdles.

c-Axis maxima in some naturally deformed quartzites may have similar origins to those produced in the experimental deformation of quartzite at relatively low temperatures and rapid strain rates (described above), in which case they can be ascribed to intragranular slip and grain rotation. For example, a deformation sequence occurs in a Central Australian metaquartzite from (i) undeformed quartz sandstone with random distribution of quartz c-axes, through (ii) rocks shortened by 20–30 per cent with c-axes distributed in a plane containing the principal axis of shortening, to (iii) rocks shortened by 65 per cent with c-axis maxima parallel to the principal axis of shortening of the rock.[63]

In the Woodroffe Thrust, Central Australia, the change from felsic gneiss to mylonite is accompanied by rotation of c-axes of original quartz grains into a girdle normal to the mylonite lineation, with symmetrical maxima at about 40° on either side of the new mylonitic foliation plane.[4, 6] These maxima are interpreted as being due to intragranular slip and grain rotation. They do not coincide with maxima produced by new recrystallised grains in the same rock. With increasing strain, only one maximum, lying close to the foliation plane of the rock, tends to develop.

A survey of the literature on the c-axis preferred orientation of quartz in mylonitic rocks ('ribbon mylonites') by Wilson and Glass has indicated that single or double maxima approximately normal to the foliation and lineation are characteristic of quartz-mylonites formed in lower greenschist facies (chlorite zone) conditions, whereas strong maxima close to the foliation and normal to the lineation are characteristic of those formed at higher grades of metamorphism.[62] Many of the observed variations from these patterns may be due to differences in slip and/or recovery processes related to variations in temperature of deformation or rate of strain.

By Recrystallisation Experiments involving recrystallisation of quartz single crystals suggest that, in intragranular recrystallisation, the c-axes of new (recrystallised) grains lie at 20–40° to the c-axis of the old grain.[28] This may be due to the effect of coincidence boundaries, as discussed in Chapter 6, although the details of the mechanism for quartz are as yet unknown. This host control is evident in annealing and stress-annealing experiments (see above), and may also be applicable to syntectonic recrystallisation, except that the results for these experiments can also be interpreted as a tendency for c-axes of new grains to lie at around 50° to λ_3, i.e. a strain control (possibly a stress control), rather than a host control.[28] The new grains are of similar size to adjacent subgrains, so that increasing rotation and misorientation of subgrains during deformation, rather than growth of subgrains, appear to have provided the high-angle boundaries for recrystallisation.

Some studies of natural single grains of quartz, that have undergone partial recrystallisation, support this host control hypothesis. For example, partial intragranular recrystallisation of quartz grains (isolated from each other by other minerals, so that host–new grain relationships can be measured relatively unambiguously) in deformed quartzofeldspathic gneisses at Broken Hill, Australia, has produced aggregates of new grains with c-axes lying between 20° and 40° of the c-axis of the original grain.[47] The orientation of the new grains is independent of regional principal stress axes inferred from the orientation of conjugate shear-zones in which the quartz-bearing rocks were deformed.

The same result has been obtained for some amphibolite facies felsic gneisses converted to mylonites in Central Australia, where the new quartz grains show a strong tendency to lie with their c-axes around 20° to that of the host grain.[4,5] The recrystallisation is inferred to have been syntectonic, and is similar to intragranular recrystallisation in syntectonic experiments (described above), except that in these naturally recrystallised aggregates fewer high angles (greater than 40°) occur and the new grains tend to be larger than the adjacent subgrains. In both the experimentally and naturally recrystallised quartz, the development of grains from subgrains seems to be the best explanation of nucleation, but some subgrain growth appears to have occurred in the natural syntectonic recrystallisation, although subgrain rotation and/or coalescence appears to have played a part in both situations. This could explain the larger size of new grains relative to parent subgrains and also the smaller range of c-axis orientations relative to the original deformed grain. The exact reason for the 20° relationship in these (and the other) metamorphic rocks is unknown. It is not obviously due to coincidence boundary effects, because the new grains are only slightly larger than adjacent subgrains.

In experimentally deformed (axially compressed) fine-grained quartz aggregates (flint and novaculite), syntectonic recrystallisation produces a variety of preferred orientations.[24] In the α-quartz stability field, c-axes show weak concentrations parallel to σ_1 at lower temperatures and higher strain rates, whereas they form a small-circle girdle about σ_1 (with r {10$\bar{1}$1} oriented preferentially parallel to σ_1) at higher temperatures and slower strain rates. There appears to be a transition between the two types of preferred orientation, and the parallelism of r and σ_1 may be due to Dauphiné twinning rather than to recrystallisation.

'Crossed-girdle' preferred orientations, similar to those in some natural quartzites, are produced in some of these experiments that involve heterogeneous bulging of the test cylinders. The preferred orientation is of nearly orthorhombic symmetry and the girdles are symmetrical with respect to λ_1 and λ_3, intersecting in λ_2. In the β-quartz stability field, similar experiments produce preferred orientations of c both parallel and normal to σ_1.

Annealing recrystallisation[24] of these syntectonically recrystallised aggregates moderately intensifies the $c \parallel \sigma_1$ preferred orientation in α-quartz, but has little effect on the $r \parallel \sigma_1$ preferred orientation. However, it nearly eliminates the $c \parallel \sigma_1$ component of aggregates showing both preferred orientations. In β-quartz, annealing intensifies the $c \parallel \sigma_1$ component, eliminating the $c \perp \sigma_1$ component; thus, annealing can produce a much stronger preferred orientation than that produced during deformation.

Mechanisms for the production of preferred orientation by recrystallisation in polygrain quartz are poorly understood (Chapter 6). At least under some conditions, intragranular recrystallisation appears to involve host/new grain crystallographic control, such that new grains in the favoured orientation grow fastest, possibly because of coincidence-boundary relationships. However, intergranular and wholesale recrystallisation probably involves several mechanisms, depending on the conditions. This is suggested by the observation that several different types of preferred orientation are produced under different experimental conditions, each being associated with different microstructures.[24] Possible mechanisms (Chapter 6) include the following: [24]

(i) *Preferred orientation due to stored strain energy.* In the deformation of polygrain α-quartz at relatively low temperatures and/or relatively fast strain rates, the rate of deformation exceeds the rate of recovery, so that strain energy is stored as dislocations. Grains with their c-axes approximately parallel to σ_1 cannot undergo basal or prismatic slip, and so should store less strain energy than grains in other orientations. Growth of these grains could account for observed preferred orientations of c parallel to σ_1 in α-quartz. This mechanism should operate during annealing until plastic strain differences between grains are eliminated, and this could explain the observed moderate strengthening of $c \neq \sigma_1$ during post-deformation annealing.

(ii) *Preferred orientation due to oriented nucleation.* The formation of nuclei in restricted orientations has been observed in the intragranular recrystallisation of a few cold-worked single crystals of quartz. However, this may be difficult to recognise in quartz aggregates, because in them recrystallisation starts at grain boundaries or at internal deformation boundaries, although it cannot be discounted as a possibility.

(iii) *Preferred orientation due to oriented growth.* If coincidence boundaries in quartz are more mobile than others, new grains growing in an oriented aggregate should grow faster through material oriented so that little atomic rearrangement is needed to change it into the new grains. In this way, new grains in favourable orientations could dominate the recrystallised aggregate. Experimental annealing of a fine, deformed α-quartz aggregate changes a broad c-axis maximum to a small-circle girdle at $35°$ to the original maximum, which could be due to preferred growth of grains with this orientation relationship to the host (i.e. similar to the relationship in intragranular recrystallisation). Experiments on β-quartz indicate that grains with their c-axes parallel to those of old grains grow fastest, which could account for the observed rapid growth of such grains in annealed deformed β-quartz aggregates described above.

Studies of natural quartzites are hampered by the difficulty of ascertaining (in a two-dimensional slice) to which old grain a particular aggregate of new grains is related. Thus, testing of the host–new grain control hypothesis is difficult. However, studies of some quartzites suggest a host control for apparent intergranular recrystallisation.[61] Furthermore, such control may explain observed random spreading or very weak preferred orientations of c-axes in some fully recrystallised natural quartzites. In these rocks pre-existing point maxima or crossed-girdle preferred orientations have been obliterated by inferred recrystallisation, as would be expected if intragranular recrystallisation produced 'cones' of c-axes at $20–40°$ from the original c-axis in each original deformed grain. Host control may also account for quartzites composed of elongate aggregates of polygonal grains with similar c-axis orientations in each aggregate, assuming large, elongate, deformed grains were present formerly. The preferred orientation would be strongest if recrystallisation occurred in the β-quartz stability field, as discussed above.

Strong preferred orientations of c-axes occur in some natural quartzites showing microstructural evidence of secondary recrystallisation, described above.[61] They consist of single point-maxima or broad girdles containing small sub-maxima, and may bear no obvious relationship to any foliation in the rock (i.e. they show no obvious control by bulk strain or mimetic growth). The secondary recrystallisation is interpreted as having commenced in initially finer-grained, polygonal, recrystallised aggregates, in which the c-axis orientation may be very weak to random. In intermediate stages of the process the quartzites show scattered c-axis concentrations in otherwise random distributions. The mechanism for the development of a strong preferred orientation in the new abnormally large grains probably involves a crystallographic relationship between the new and old grains, but this is difficult to visualise in an initially random aggregate, in which coincidence boundaries should be randomly distributed. Possibly the initial aggregate has an optically undetectable preferred orientation needing X-ray diffraction, but whatever the explanation, this problem merits experimental investigation.

Variation in Quartz Microfabric with Metamorphic Conditions

Wilson has investigated changes in quartz microfabric (microstructure and c-axis preferred orientation) in metaquartzites from a prograde regional metamorphic sequence near Mount Isa, Australia.[61] The changes are:

(i) original detrital grains become flattened and elongated in a foliation, and show deformation bands, undulose extinction and serrated grain boundaries in the chlorite zone;
(ii) small, optically strain-free grains grow with their c-axes controlled by the old host grains in the chlorite-biotite zone;
(iii) these new grains grow and eliminate old grains (biotite zone), forming a polygonal microstructure (biotite-cordierite zone);
(iv) exaggerated growth (probably secondary recrystallisation, as discussed above) of a few grains in the polygonal aggregate results in a coarse-grained mosaic of grains with complex shapes in three dimensions (sillimanite zone).

In the least deformed quartzites the quartz c-axis patterns are random, but, with increasing deformation and metamorphic grade, they change successively to peripheral girdles, crossed girdles, random strong girdles and point maxima. These changes may be due to a progressive sequence of changes in deformational/recovery conditions accompanying progressive metamorphism, or, alternatively, may be contemporaneous, in which case they could result from different physical conditions in different parts of the same rock sequence during one deformation (and presumably heating) episode. The problem is closely related to the general problem of progressive metamorphism discussed in Chapter 2, namely: do the mineral assemblages of each zone pass successively through those of all lower grade zones, or are all assemblages formed at the same time in response to a gradient in physical conditions across the area concerned?

Another complication in this area[61] is that the pattern of preferred orientation in the quartz of a particular metamorphic grade depends to a variable extent on the proportion of accessory minerals present, especially mica. This is less evident at lower grades, but mica grains can severely restrict quartz growth in the coarsening stage (Chapter 5), thereby preserving patterns of preferred orientation characteristic of lower grades in mica-poor rocks.

Bell observed marked differences in recrystallisation microstructures between gneisses of the amphibolite facies and those of the granulite facies deformed on either side of the Woodroffe Thrust, Central Australia.[4] On the amphibolite facies side, the size of subgrains and new grains in quartz, the degree of recrystallisation and the preferred orientation are much greater, relative to strain, than on the granulite facies side. Moreover, new grains nucleate on host grain edges and deformation band boundaries on the granulite facies side, whereas they are confined to host grain edges on the amphibolite facies side. Much more pronounced subgrain rotation appears to have occurred on the amphibolite facies side. In addition, on the granulite facies side the host/new grain crystallographic relationships are strong (the *c*-axes of new grains lying around 20° to that of the host, as in most experiments on quartz recrystallisation), whereas on the amphibolite facies side the new grains have a nearly uniform angular distribution about the *c*-axis of the host. All these features suggest differences in nucleation mechanisms and in slip and climb systems from the amphibolite to the granulite facies rocks. This would be related to the type and/or generation and/or movement of dislocations, which could well be controlled by the higher strain rate and/or water content (which are likely to be related in silicates) in the amphibolite facies gneisses.

Stress Indicators in Deformed Quartz[3, 11]

Sub-basal I lamellae (Fig. 7.1) can be used as potential indicators of stress axes, provided the deformation involves pure shear, or else of the last incremental stress imprinted on the rock in a more general deformation. In the experimental deformation of quartzite, deformation lamellae form at angles near, but generally less than 45° to the experimental σ_1 direction. Also, poles to lamellae lie nearer σ_3 than do c-axes in the same grains, and c-axes lie closer to σ_1 in more highly deformed parts of grains (where lamellae are profuse) than do those in less deformed parts. However, for this information to be useful in the interpretation of naturally deformed quartz aggregates, the deformation must have been homogeneous on the scale of the sample measured, and we must be certain that the natural lamellae are identical to, and were formed in the same way as, the experimental lamellae, which may require transmission electron microscopy for verification. For example, examination of some naturally deformed quartz in the electron microscope has shown that, although some deformation lamellae (especially those formed in environments of inferred rapid strain rate) lie in slip planes and so may be correlated with the experimental lamellae, most natural lamellae are dislocation walls forming narrow basal subgrains; their present position and attitude are the result of recovery, not slip, and so they cannot be correlated with the experimental lamellae.[60]

OLIVINE

Introduction

Olivine in most terrestrial rocks is magnesium-rich and commonly shows evidence of deformation, recovery and recrystallisation in the optical microscope. Geological evidence indicates that most olivine-rich rocks (peridotites and dunites) are transferred from the earth's mantle to its crust as accidental fragments in basalt magma and as intrusions of the alpine type. If some or all of the deformation features shown by the olivine were produced in the earth's mantle, we must investigate the deformation and recrystallisation of magnesian olivine at high temperatures, high confining pressures and slow strain rates in order to infer the mechanisms involved. Detailed studies of natural olivine-rich rocks[1,38,40,41] have revealed considerable complexity of microstructures and preferred orientations, but we will concentrate on the main general features shown by experiments and natural rocks.

Experimental Deformation,
Recovery and Recrystallisation of Olivine[7,11,12,25,41,44,46]

Experiments show that optically determined slip systems in olivine vary with temperature, confining pressure and strain rate (Figs 7.2, 7.3).[11,12,46]

Electron microscopic examination of dislocations in experimentally deformed olivine has confirmed these systems.[7,25] For example, dunite deformed at 800°C has mainly dislocations with **b**=[001], whereas an olivine single crystal deformed at 900°C has mainly dislocations with **b**=[100] although both types of dislocation occur in each sample, show-

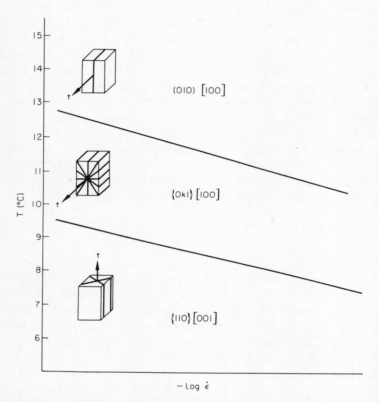

Fig. 7.2   Slip systems in experimentally deformed olivine aggregates as functions of temperature and strain rate, at constant confining pressure of approximately 15 kb. After Carter and Ave'Lallemant.[12]

ing that slip occurs in both directions above and below the transition temperature. Therefore, the transition appears to involve a change in the relative contributions of each slip system.[7] A ten-fold decrease in strain rate lowers the temperatures of the transitions between slip systems by about 50°C, but confining pressure has a much smaller effect, in contrast to quartz.[11] Slip in the direction [100] predominates at higher temperatures and slower strain rates, and almost all deformed terrestrial olivine shows optical evidence of such slip.

Microstructures indicating heterogeneous deformation by slip are characteristic of experimental deformation at relatively lower temperatures.[11,46] They consist of slip bands (not visible in transmitted light), deformation lamellae and kink bands (some varieties of so-called 'un-

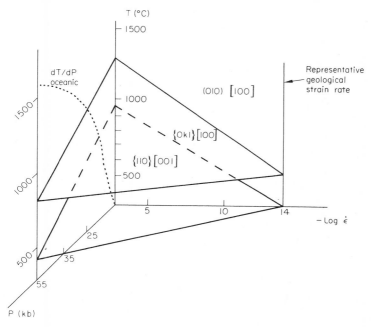

Fig. 7.3 Slip systems in olivine as functions of temperature, confining pressure and strain rate, extrapolated from experimental data to geologically reasonable strain rates and to high confining pressures. After Carter, p. 5521, © by American Geophysical Union.[11]

dulatory extinction'). The *deformation lamellae* are optically identical to those in quartz, and have been shown by electron microscopy to consist of elongate tangles of dislocations formed as a result of work-hardening.[25] They form parallel to slip bands and at orientations of high resolved shear stress and, therefore, are considered to lie in active slip planes. At low to moderate temperatures they are parallel to (100) and planes of the form {110}, and at temperatures above 1 000°C (in experiments at strain rates of around 10^{-5} sec^{-1}) they are parallel to many planes of the type (0kl); i.e. they are in the zone [100]. The *kink bands* are produced in deformation experiments under all or most conditions, but are common only where the macroscopic longitudinal strain is greater than about 4 per cent. The external rotations across kink bands in olivine generally are less than 10°, and the bands are parallel to (100), (001) and (010). Electron microscopy[25] shows that the kink band boundaries at small strains consist of parallel arrays of unit edge dislocations with Burgers vector=a[100], separated by areas of low dislocation density. The dislocations are not all strictly confined to the one plane. At larger strains the dislocation density in the walls increases and the misorientation across the kink boundaries also increases.

Dense dislocation tangles occur in forsterite single crystals deformed at 800°C and 1 000°C (and 10 kb), the tangles forming the walls of a cell structure.[44] They are probably formed by jogging of the common dislocations with b=[001] by other intersecting dislocations, although some cross-slip may have occurred also.[44]

In deformation experiments above 1 000°C no deformation lamellae are produced, the only optical evidence of plastic deformation being kink bands and subgrain boundaries. The absence of deformation lamellae is believed to be due to many dislocations climbing out of their slip planes, leaving too few to give an optical image of the plane. Cross-slip and climb of dislocations is marked in deformed forsterite crystals deformed at 1 000°C and 10 kb.

The effect of variation in the amount of strain on olivine deformation is shown by axial compression experiments[41] on polygrain olivine, at 1200–1300°C, 13–15 kb confining pressure and strain rates of about 10^{-5}/sec. These produced bodily rotation of initially inequant grains at low strains (less than 30 per cent shortening), accompanied by limited slip (associated with kinking in suitably oriented grains). This causes the shortest dimension [010] of the grains to rotate progressively closer to the experimental σ_1 direction, and [100] and [001] to migrate towards the plane normal to σ_1. At strains larger than 30 per cent, slip and recrystallisation (see later) occur, the slip (mainly (010 [100]) apparently being responsible for most of the preferred orientation produced, namely an intensification of the preferred orientation started at smaller strains. Kink bands are common, their boundaries being normal to the flattening plane. Some grain boundaries become intensely deformed and obscured by the production of fine, syntectonically recrystallised grains (see below), which grow larger if the aggregate is annealed or if the experiments are carried out at strain rates of 10^{-6} sec^{-1}. The microstructures produced are very similar to those produced in the experimental deformation of quartzite.[52, 53] Remnant augen occur in orientations unsuitable for slip in the experimental stress field (as for augen in experimentally deformed quartzite). At strains larger than 40 per cent, syntectonic recrystallisation becomes important. In experiments at high temperatures and small strain rates, recrystallisation may become the predominant mechanism of deformation.[2]

Electron microscopy of samples showing extensive evidence of dynamic recovery and recrystallisation reveals a low dislocation density, helical dislocations, dislocation loops and extensive stabilised networks of dislocations forming subgrain and (recrystallised) grain boundaries. These features indicate climb of edge dislocations and cross-slip of screw dislocations.[25, 44] Recrystallisation begins at grain boundaries, and then partly replaces deformed grain interiors.[2] Experimentally, it has been found that the recrystallisation temperature is lowered by about $50°C$ for a ten-fold reduction in strain rate, so that recrystallisation could occur at temperatures as low as $500°C$ at geologically reasonable strain rates of 10^{-14} sec^{-1}.[2]

Electron microscopy has indicated that the new recrystallised grains show the same strain effects as unrecrystallised grains in both experimental and natural olivine aggregates. This confirms that recrystallisation probably accompanied deformation (i.e. it was syntectonic, rather than static).[25] The presence of water (released by dehydration of talc jackets in the experiments) enhances diffusion[26] and produces larger recrystallised grains, but has no noticeable effect on the initial recrystallisation temperature.[2]

In some experiments at higher temperatures or lower strain rates, olivine aggregates completely recrystallise to finer-grained, optically strain-free grains with irregular to smooth boundaries, the grains generally being slightly flattened in the experimental σ_2–σ_3 plane. This produces a foliation similar to that shown by some naturally deformed olivine-rich rocks.[2]

Natural Deformation, Recovery and Recrystallisation of Olivine

Optical and electron microscopic examination of natural alpine and most xenolithic peridotites[25] indicates that olivine deformed in the earth's upper mantle flows by the same mechanisms as in high-temperature experimental deformation, mainly by slip of dislocations with **b**=[100], and simultaneous dynamic recovery producing kink bands and prismatic subgrains. Recrystallisation is common, the microstructures produced being similar to those formed by dynamic recovery and recrystallisation in experiments carried out at PT conditions inferred by chemical evidence to be applicable to the origin of these rocks.[2, 41] For example, both deformed relics and new recrystallised grains are common in alpine peridotites,[41] many olivine-rich xenoliths[41] and dunite-mylonites.[8] In some of these rocks the recrystallisation appears to be of high-temperature, static origin, since the low dislocation density within the recrystallised grains is similar to that in olivine believed to have statically recrystallised in peridotite xenoliths immersed in liquid basalt.[44] Inferred annealing of mantle peridotite xenoliths in hot basalt promotes static recovery, the olivine grains retaining their major deformation/recrystallisation features, but losing their generally distributed dislocations.

Some alpine peridotites show the effect of later lower temperature and/or more rapid deformation that may have occurred during or after relatively 'cold' emplacement in the crust. Olivine inferred to have been deformed at rapid strain rates (e.g. in chondritic meteorites) shows optical evidence of [001] slip. Both optical evidence and electron microscopic examination of dislocations indicate that [001] slip also has occurred in the olivine of peridotite xenoliths in kimberlite intrusions, but some people attribute this to rapid strain rates (associated with the intrusion of the kimberlite) and others[42] to slow strain rates (associated with flow in the earth's mantle). The dislocation systems in some naturally deformed olivine inferred to have been derived from the mantle are more complex than in experimentally deformed olivine.[42] For example, deformed relics in certain dunite-mylonites contain subgrain boundaries parallel to (100), (001), (010) and (110), with evidence of dislocations with Burgers vectors parallel to [110], [010], [001], and [11$\bar{2}$], and also networks of intersecting dislocations.[8] This complexity has been attributed to either the very slow rates of mantle deformation or the complex deformation/recovery history of the rocks concerned.

Preferred Orientation of Olivine

(i) *By deformation.*[41] As discussed previously, bodily rotation of inequant grains (if present) and, especially, intragranular slip (with accompanying kinking and grain boundary deformation) can produce preferred orientations of [010] parallel to the direction of shortening, and [100] and [001] lying in the plane of flattening. This general kind of preferred orientation is common in alpine and xenolithic olivine-rich rocks,[1,2,40,41] which may indicate that similar orienting processes have occurred in the earth's mantle. Some [100] or [001] concentrations may occur in natural aggregates.[2]

However, the detailed mechanism is in doubt, because a strong predominance of dislocations with Burgers vectors of a[100] severely restricts the production of preferred orientation by slip alone, although a few reports exist of dislocations with other Burgers vectors in some naturally deformed olivine (see above). The commonly observed kinking could assist the process and grain-boundary sliding also may have occurred, especially since intergranular recrystallisation is common.

(ii) *By recrystallisation.*[2] Intragranular recrystallisation in some experiments produces a preferred orientation of new grains related to the orientation of the original grains in a manner similar to most recrystallisation in single crystals of quartz. However, in intergranular recrystallisation the new grain orientations are symmetrically related to the experimental principal stresses. As grain-boundary recrystallisation advances, the old grains become consumed, and grains oriented favourably in the stress field grow preferentially. The preferred orientation in completely recrystallised aggregates tends to be: [010] approximately parallel to the experimental σ_1 direction (also the direction of shortening) and [001] and [100] lying in girdles parallel to $\sigma_2=\sigma_3$ (the flattening plane).[2]

So, identical patterns of preferred orientation are produced by inferred slip processes and by inferred recrystallisation processes in experiments. The latter appear to be favoured at smaller strain rates and, possibly, higher temperatures.[41] Probably both processes contribute to preferred orientations in natural olivine aggregates,[40, 41] but their relative importance is a matter of debate at this stage; it probably varies from one rock to another. One point is clear, however, namely that motion of dislocations, and not of point-defects, dominates the flow and recrystallisation of olivine. Apart from theoretical objections to large-scale flow by migration of point-defects,[8] the characteristic arrangement of line-defects in all the natural and experimental samples shows the predominance of mechanisms involving dislocations.[7, 8, 25, 42, 44]

LAYER SILICATES

Introduction

Layer silicates are characterised by a high degree of structural anisotropy, which introduces special difficulties into the understanding of their mechanical and recrystallisation behaviour. The problem of developing preferred orientations and foliations in aggregates of layer silicates is a long-standing one, and controversy over the origin of cleavage in slates has raged for over 150 years.

Deformation of Mica Single Crystals[17, 19]

Direct observations of dislocations and deformation experiments on biotite single crystals have indicated that [100], [110] and [1$\bar{1}$0] are slip directions in the basal (001) plane, as predicted by the fact that these are the three directions of closest packing in the basal plane. However, insufficient experiments, involving deformation of single crystals in various orientations, have been done to be able to say that these are the only slip systems in biotite. Furthermore, we cannot assume that all intragranular slip in layer silicates is controlled by motion of dislocations; some may be due to cataclastic failure along basal planes, as suggested by opening up of cleavage planes, both in natural rocks (Paul Williams, Stan White, personal communication) and some experiments.[17]

Heterogeneous deformation is typical in experiments, the main microstructural features of deformed single crystals of biotite and muscovite being kinks that are very variable in size, orientation, abundance and shape. At low temperatures (300–500°C), kinks are narrow, abundant and at high angles to the direction of shortening, whereas at higher temperatures (600–700°C) they are broader, less numerous, at lower angles to the direction of shortening and with more total area of undeformed biotite between them.[17]

The kink boundaries generally are optically sharp, but where the angle of bending (ω) is less than 45°, they become less well defined, passing laterally into regions of uniform bending of (001). Careful optical examination has shown that the kink boundaries are about 1–2 μm wide, so that, if they consist of dislocations, their dislocation structure may be relatively complex.[17] Alternatively, many kink boundaries (especially high-angle ones) may be fractures.

Deformation of Fine-Grained Layer Silicate Aggregates[17,36,54]

Hydrostatic compression of fine, polygrain hydroxyphlogopite, talc and brucite (grown hydrothermally from their constituent oxides at 300–600°C and 3–5 kb confining pressure) has produced random orientations. Axial compression of these layer silicate aggregates, both at room temperature and high temperature, has produced a moderate preferred orientation of (001) planes normal to the axis of shortening (λ_3).[36]

Deformation of finely ground 2M muscovite at about 700°C and of fine-grained phlogopite at about 700–900°C (all at about 15 kb confining pressure and a strain rate of around 10^{-4} sec^{-1}) has produced the same preferred orientation of (001) planes, although the grain-size remains very small (up to 1 μm).[54] The mica (001) planes always show a strong tendency to become aligned normal to the axis of maximum shortening (λ_3), even where rotational strain components are involved. In addition, the concentration of poles to (001) parallel to λ_2 is greater than that parallel to λ_1. The intensity of the preferred orientation increases with increasing strain. Where the strain can be shown to be pure dilation ('hydrostatic deformation'), the mica grains are random.

The same general result has been obtained in experiments deforming fine synthetic fluorphlogopite at room temperature and pressure,[54] fine 2M muscovite at 200°C and 15 kb confining pressure[54] and fine talc at 25–700°C and 3–4 kb confining pressure,[17] most of these temperatures being far too low for any recrystallisation to have occurred, so that the preferred orientation must be of mechanical origin. Deformation at room temperature and 3 kb confining pressure of fine synthetic hydroxyphlogopite has produced much lower preferred orientations than those produced in the other experiments at high temperatures.[17] Deformation of a compacted mixture of synthetic fluorphlogopite and common salt at room temperature at confining pressures up to 2·07 kb and strain rates from 10^{-4} to 10^{-6} sec^{-1} has caused reduction in mica grain-size, warping of mica grains and bodily rotation of mica grains into zones of fault-like displacement through the deformed samples.[37] The deformation appears to have been due to a combination of intragranular and intergranular processes.

Deformation of fine-grained synthetic hydroxyphlogopite grown from an oxide-carbonate mix with an interstitial fluid has been achieved by axial compression at strain rates of about 10^{-4} sec^{-1}, a confining pressure of 3 kb and a temperature of 500°C.[17,20] In central areas of the capsules where most reaction occurs, the yield, grain-size and preferred orientation of phlogopite are high, these areas enlarging with increasing strain. Unreacted particles generally occur as flattened ellipsoidal grains or aggregates, some of the flattening being due to the making of the pellet, the remainder being due to deformation in the experiments. The larger phlogopite grains produced are 1·5 mm long (parallel to (001)) by 0·2 mm wide, and their shapes commonly are controlled by the non-micaceous particles, even to the extent that thin phlogopite grains locally wind between them. The large phlogopite grains grow very quickly and have a very strong preferred orientation normal to the shortening direction. Their preferred orientation and length-to-width ratios increase with increasing strain. Apart from local weak bending around non-micaceous fragments and some possible fine kinks in a few grains, the large phlogopite grains show surprisingly little internal deformation (even for 40 per cent bulk longitudinal strain), considering the potential instability of long, thin grains in a strongly deforming matrix. The preferred orientation of the finer-grained (less than 0·2 mm long) phlogopite is much lower than that of the coarser phlogopite, being controlled by the shapes of the non-micaceous grains.

Experiments of the same type, but designed to vary the pore pressure, show that, with increasing pore pressure, the yield, grain-size and degree of preferred orientation of phlogopite decrease, so much so that, if the pore pressure equals the confining pressure, low yields, low pre-ferred orientations and no coarse-grained mica are produced.[17] Possibly the low yields of phlogopite can be explained by the influx of argon (by which the increase in pore pressure was accomplished), which should dilute the water present (thereby reducing f_{H_2O}) and restrict the production of phlogopite (Chapter 2).

Recovery and Recrystallisation of Mica Single Crystals[17, 18]

Single crystals of phlogopite deformed at 300°C and 10 kb confining pressure up to 35 per cent longitudinal strain recrystallise if annealed for 5–25 minutes at temperatures greater than about 1 050°C. No re-crystallisation occurs if annealing is carried out below this temperature, even for four hours; this applies to biotite and muscovite also. Attempts to recrystallise deformed coarse-grained talc at 650°C for two hours have been unsuccessful. Three stages appear to comprise the inferred recrystallisation process in phlogopite.

(i) *Inferred migration of kink boundaries*, producing serrations parallel to (001) planes in either of the kink limbs. With increasing temperature or time, the serrations become larger and spread along the kink boundaries. Eventually kink bands may disappear, leaving small bleb-like regions, from which new grains may grow with their (001) planes very close to those in the shrinking kink band. Kink boundaries with $\omega = 70° - 80°$ are the most mobile.

(ii) *Nucleation and rapid growth of large new grains*, commonly with (001) oriented similarly to that of the adjacent deformed mica, eventually producing a 'decussate' (rational-impingement) type of fully recrystallised aggregate, through parts of which former kink band boundaries can be traced with preserved (001) serrations. However, some grains cut across the aggregate with no apparent crystallographic relationship to the deformed material.

(iii) *Nucleation and growth of smaller new grains in the larger recrystal-lised grains*, as well as along some low-ω kink boundaries, and also scattered through the deformed part of the crystal adjacent to the hotter zones. The larger of these grains are rectangular, with (001) planes where measurable, but no systematic relationship to either the orienta-tion of the host crystal or the bulk strain has been detected.

Electron microprobe analysis of experimentally recrystallised phlogo-pite shows systematic chemical differences (mainly in Na, K and total

Fe) between the new recrystallised grains and the old grains.[17, 18] Chemical differences (mainly in Ti, Al, Mg and Fe) also exist between new and old grains in a naturally recrystallised biotite.[17, 18]

Observations of some natural partly recrystallised biotites have suggested various processes, namely: (i) evidence of local kink boundary migration, and recrystallisation concentrated along kink boundaries, the (001) planes of new grains tending to be parallel to the kink boundaries, and the new grains tending to be longer where their (001) planes lie at low angles to the kink boundaries;[4, 17] (ii) apparent extensive migration of parts of kink boundaries, producing rational-impingement aggregates (Paul Williams, personal communication); (iii) inferred general recrystallisation through most of a biotite grain without many kinks remaining (although recrystallisation generally is closely associated with those few kinks that are present) and without evidence of kink boundary migration;[17] or (iv) recrystallisation at mica grain boundaries, the new grains being oblique to (001) of the host.[4]

A number of nucleation mechanisms have been suggested to explain the above observations, namely:

(i) *random or homogeneous nucleation*, due to random thermal fluctuations in a uniformly deformed grain, which is unlikely, as calculations based on the known approximate surface energy of mica indicate that the size of a critical nucleus would be far too large;[17]

(ii) *heterogeneous nucleation* on kink boundaries (whether fractures or not), and possibly on deformed grain boundaries, and on parted cleavage surfaces (if present in the deformed material);

(iii) *pre-formed or subgrain nucleation*, which also is unlikely, as both the formation and rotation of subgrains in mica are not favoured by the restricted dislocation structure, although closely associated grains and apparent subgrains (with a gradual increase in one at the expense of the other) in some deformed rocks suggest subgrain rotation during syntectonic recrystallisation;[4]

(iv) *bulge nucleation*, which could account for the observation of kink serrations, and possibly some rational-impingement aggregates, but, where grains have orientations very different from that of the deformed material, this mechanism is not applicable;

(v) *nucleation involving release of chemical free energy*,[17, 18] which does not depend on the release of stored strain energy alone, although this could localise and assist nucleation in regions of high heterogeneous strain. This mechanism is supported by the observation of chemical differences between old and new grains,[17,18] and could help to explain the perplexing fact that, although experimental recrystallisation of mica single crystals is extremely difficult, natural aggregates inferred to have been formed by the recrystallisation of formerly larger grains are common in metamorphic rocks. Perhaps these aggregates grew during chemical reactions that provided a strong driving force for recrystallisation. Of course, this is not meant to imply that all 'decussate' mica aggregates in metamorphic rocks formed by the recrystallisation of formerly large mica grains.

Experimental 'Recrystallisation' of Fine-Grained Layer Silicate Aggregates

Increase in the grain-size of compacted fine-grained (43–61 μm), synthetic fluorphlogopite[54] has been achieved under anhydrous conditions at 1 250–1 300°C. In one experiment involving hydrostatic annealing for 9·5 hours at 1 300°C and 15 kb confining pressure, a typical 'decussate' aggregate of strain-free grains with distinct (001) boundaries was produced. Their maximum length parallel to (001) in section was 150 μm, and their length-to-thickness ratio ('aspect ratio') ranged from 2 to 8 (averaging 4). Some syntectonic growth also appears to have occurred in the deformation of fine synthetic hydroxyphlogopite at only 500°C, as described previously,[17] and growth of fine talc has been produced in syntectonic and post-deformation annealing experiments above 600°C.[17] In the talc experiments, the preferred orientation increases with increasing strain, and growth of grains occurs only in deformed parts of the sample. Where they are large enough to examine optically, most talc grains with (001) originally at low angles to λ_3 are strongly kinked, whereas those at moderate to high angles to λ_3 are undeformed.

These experiments suggest that growth of small grains in deformed aggregates of layer silicates is promoted by: (i) large amounts of intragranular strain (since strongly deformed grains contain many kinks that may have dislocation concentrations along their boundaries, if ductile, or may have large exposed surfaces, if brittle); (ii) large grain-boundary area (which could be related to the initial small grain-size, and possible grain-size reduction by intragranular cataclasis)[36]; and (ii) high-energy dislocation arrangements in deformed edges of grains.[17]

The increase in grain-size tends to be accompanied by an increase in preferred orientation, such that in talc experiments, for example, grains with (001) at low to moderate angles to λ_3 disappear quite suddenly. No evidence of nucleation and growth of new grains inside old grains has been detected.[17]

In these experiments, it is not clear whether growth and adjustment of grains has occurred as (*a*) a process involving sintering of fragments, (*b*) a process driven by strain energy in deformed grains, (c) growth from an interstitial fluid phase (part of which could have been released from the original partly hydrous oxide mix), or (d) a combination of these factors.

The preferred orientations produced in these experiments are of the same kind as, though sometimes more intense than, those produced by the deformation of fine aggregates without growth of grains, and this applies both to static annealing and syntectonic experiments. So, it appears that mechanical rotation may have made a major, if not predominant, contribution to the preferred orientation.

Preferred Orientation Mechanisms in Layer Silicates[17, 36, 54]

Most reported preferred orientations of layer silicates involve alignment of (001). However, preferred orientations of [100] and [010] also have been reported.

The hypotheses that have been advanced to explain (001) preferred orientation may be grouped, as follows:

(i) mechanical rotation of relatively rigid plates into the $\lambda_1 \lambda_2$ plane, with or without rotation due to intracrystalline slip;

(ii) mechanical rotation of plates into a plane of high shear strain, which may form at various angles to λ_3 or σ_1, and may rotate towards one or both of them during the deformation;

(iii) growth of grains with their (001) planes normal to either λ_3 or σ_1, or parallel to a pre-existing structure in the rock (=mimetic growth), or both.

Therefore, the following general mechanisms need evaluation:

(a) *Rotation due to inequant shape.* Calculations enable prediction of the degree of preferred orientation to be expected if rotation of rigid, isolated tabular bodies (such as mica plates) occurs in a perfectly homogeneous deformation. Unfortunately, mutual interference of grains would occur in actual rock deformation, and this effect has not yet been calculated successfully. The kind of (001) preferred orientation produced by mechanical rotation due to shape alone is similar to that observed in naturally and experimentally deformed layer silicates. This mechanism also can explain preferred orientations of [100] or [010], but only provided the grains are elongate parallel to these directions.

(b) *Rotation due to intragranular slip* should tend to cause slip planes to rotate so that they become normal to the direction of maximum shortening. Either grain segments in kink bands would rotate from low to high angles to this direction, or more simple bending by (001) slip could cause parts of grains to rotate. General predictions for the axial compression of an aggregate of grains with only one slip plane are that the orientation of this plane normal to λ_3 should occur only after very large strains.[17] However, large strains are possible in rocks, so that this may be effective, in conjunction with intergranular slip, especially if fracturing and slip without cohesion along (001) planes occurs.

Intragranular slip also has been invoked to explain preferred orientations of [010] or [100], namely by slip on (001) parallel to the direction concerned, accompanied by grain rotation causing preferred orientation of (001). However, because experiments show an equal preference for slip on six possible directions (i.e. two senses of slip in each direction) in (001), no preferred orientation can be caused by (001) slip. Therefore, any preferred orientation of this kind must be due to another mechanism, such as growth, recrystallisation or rotation of initially inequant grains.

(c) *Preferred orientation due to differential strain energy* can involve:

(i) *stored energy of deformation*, such that the most deformed grains in an aggregate have higher internal energy, and so may tend to recrystallise more readily than less deformed grains (as discussed previously, and in Chapter 6), thus producing a preferred orientation. For example, selective removal of kinked mica grains (those with (001) at low angles to λ_3) would produce a preferred orientation similar to that observed.[17] This could occur either by migration of boundaries from undeformed grains (with (001) close to λ_3) into adjacent kinked grains (i.e. solid state replacement), or preferential solution of kinked grains in the fluid phase (assuming it is present in the experiments) and redeposition on undeformed grains lying with (001) close to λ_3.

(ii) *elastic energy due to the applied stress*, which depends on the orientation of an anisotropic crystal in the stress field. Thermodynamically, the differences in strain energy between different possible orientations could cause a preferred orientation to develop, as material diffuses from some parts of grains to others, preferably in a fluid phase. However, it is difficult to prove that such a mechanism actually takes place.

Means and Paterson have suggested that, in their experiments on the deformation of synthetic layer silicates, the free energy change associated with forming the layer silicates from their constituent oxides is very large relative to possible free energy differences associated with different orientations of stress (as in (ii) above) or plastic strain (as in (i) above).[36] This could mean that the large silicates probably grow quickly, in random orientation, when the temperature is first raised, and deformation involves subsequent mechanical rotation of the grains, probably accompanied by some intragranular strain (especially around grain-boundaries) where free rotation is impeded. The implication also is that the rates of recrystallisation of deformed grains are negligibly small in these experiments, compared to initial growth rates and mechanical rotation rates. We should keep these conclusions in mind for geological situations involving the formation of layer silicates in metamorphic reactions with large free energy changes, such as in the development of slaty cleavage during the initial metamorphism of clay-rich aggregates.

(*d*) *Preferred orientation due to recrystallisation.* As described above, some natural biotite contains smaller new grains that apparently re-crystallised with their (001) planes parallel to kink boundaries. Since these boundaries tend to lie at high angles to λ_3, such inferred recrystal-lisation can cause a preferred orientation of (001) approximately normal to λ_3. However, whether this process is of widespread applicability is unknown.

As reported earlier, annealing and syntectonic heating of fine-grained talc causes an increase in grain-size and a relatively sudden elimination of grains (commonly kinked) with (001) at low angles to λ_3, with the con-comitant production of unstrained grains with (001) at high angles to λ_3. However, whether this is a recrystallisation process or simply the growth of mechanically oriented grains and/or fragments is unknown at this stage.

(*e*) *Preferred growth due to constraints* probably occurs in the experi-ments (described previously) involving deformation of fine hydroxy-phlogopite grown from an oxide-carbonate mix with an interstitial fluid phase.[17] Mechanical rotation of grains, acting alone, cannot explain the observations made on this deformed material, although it may have con-tributed to the preferred orientation, especially of the small phlogopite grains. Surprisingly, increasing the pore fluid pressure reduces the pre-ferred orientation of large phlogopite grains, whereas intuitively, it might have been expected to force grains apart allowing mechanical rotation of mica flakes into parallel alignment. This, together with the lack of internal deformation, the ragged shapes and the temperature-strain dependence of the yield and preferred orientation of large phlogopite grains, all points to grain growth constrained into a preferred orienta-tion during deformation. In a sense this is mimetic growth, because the flattening of unreacted grains during initial compaction produces (*a*) *channels of high permeability* (Fig. 7.4), which promote anisotropic distri-bution and supply of components for mica growth, and (*b*) *anisotropic growth constraints*, owing to the shapes of solid grains adjacent to the growing mica crystals. These factors, combined with the known strong tendency of mica to grow fastest parallel to (001), can explain the high degree of (mimetic) preferred orientation produced. Such a mechanism may be relevant to natural situations in which layer silicates grow during deformation in the presence of a fluid phase. It can be reconciled with

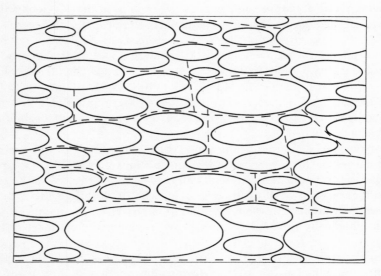

Fig. 7.4   Idealised sketch of an aggregate of compacted ellipsoidal particles of various sizes, nearly touching. Continuous channels are much longer parallel to the preferred orientation of ellipsoids than at even a small angle to it. The channels are very short at large angles to it. After Etheridge.[17]

'dewatering' hypotheses for the initiation of slaty cleavage.[35, 45] In addition to producing a mechanical alignment of any layer silicates that remain stable under the metamorphic conditions involved, dewatering could produce anisotropic channels by flattening and/or alignment of non-micaceous grains. Layer silicates precipitating from a fluid phase under these constraints should produce a strong preferred orientation, according to the above experiments.

(f) *Preferred orientation owing to growth in a temperature gradient* has taken place in some experiments on synthetic fluorphlogopite in which melting occurred.[17] Large (up to 1·3 mm), long (up to 30 in aspect ratio) crystals grew with a strong preferred orientation of (001) normal to the sample wall on which they nucleated. They appear to have grown during a 30-second quench from 1 300°C to 100°C at the termination of the experiments. Such experiments may have some relevance to layer silicates growing in veins or pegmatitic bodies.

Systematic Study of Natural Deformation and Recrystallisation of Mica

In a transition from felsic gneiss to fine-grained mylonite in Central Australia, the biotite initially is coarse-grained and shows strain shadowing with some kinking in the earliest stages of deformation.[4,6] The kinks have $\omega = 25-90°$, the higher mismatched boundaries commonly being serrated, or the sites for the development of new biotite and quartz grains. The new biotite grains mostly are at a small angle to the kink boundaries, which themselves tend to lie parallel to the developing foliation in the rock. This alignment of kinks may be due to slip on (001), accompanied by bodily grain rotation (controlled by the strain in surrounding grains of framework silicates). The combination of aligned kinks and recrystallisation of new grains with their (001) planes parallel to the kinks results in a preferred orientation of (001) parallel to the foliation.

As the foliation becomes more strongly developed, most of the large biotite grains are bent or kinked, only those initially at a small angle to the new foliation remaining unaffected. New grains (presumably formed by recrystallisation) are common around the edges of most biotite grains, even undeformed ones. The higher angle kink boundaries are commonly either serrated or are the sites of recrystallisation. Kinking and marginal recrystallisation cause elongation of the biotite aggregates parallel to the new mylonitic foliation. In the mylonite proper, all old mica grains disappear.

More systematic investigations of this type are needed, preferably coupled with chemical analysis of new and old grains, and an attempt to relate deformation to metamorphic reactions (Chapter 8), which should be important in the recrystallisation of compositionally complex minerals such as layer silicates (Chapter 6).

PLAGIOCLASE

Introduction

Little is known about deformation and recovery mechanisms in feldspars, despite their abundance in rocks. Their atomic structure is complicated and electron microscopy and X-ray diffraction have shown that most of them consist of submicroscopic mixtures of two or more feldspar phases. So, detailed atomic mechanisms will be difficult to work out.

Feldspar behaves differently to quartz and layer silicates in most deformed rocks. Generally it tends to be stronger. For example, in many mylonitic rocks formed by the deformation of mica-bearing quartzofeldspathic rocks, the quartz is very elongate and extensively recrystallised, and the mica occurs in elongate, decussate (apparently recrystallised) aggregates, but the feldspar remains as large grains or fragments that may or may not show internal deformation and/or minor recrystallisation.

However, in some rocks, feldspar appears to have undergone extensive deformation and recrystallisation, most work so far having been done on plagioclase. Plagioclase recrystallisation is very well shown by certain rocks deformed under granulite facies conditions, where increased temperatures probably promote dislocation movements.[29,39,57] None the less, recrystallised plagioclase may also occur in some greenschist facies rocks, so that metamorphic grade cannot be the only controlling factor.

Experimental Deformation of Plagioclase[9,10]

More experimental work needs to be done on plagioclase deformation, but available results indicate the following:

(i) Most experimental deformation occurs by twinning on both the albite or pericline laws in plagioclases in the compositional range $An_{30}-An_{96}$ (possibly An_{100}) at 800°C and 10 kb confining pressure. No microscopic twinning occurs in experiments below 800°C. Twins of both laws may occur together or separately in different parts of the same grain, owing to deformational heterogeneities.

(ii) In the experimental deformation of ordered (low temperature) albite and a peristerite (An_{13}), slip occurs on (010) parallel to an irrational direction that probably is the glide direction for albite twinning. However, grains oriented favourably for slip do not twin, and vice versa. The slip is accompanied by kinking, involving a rotation of up to 7°, but kink band boundaries are not well developed. Slip on (010) may occur in plagioclase of other compositions, but suitable experiments have not yet been carried out. Slip on (001) has not yet been observed, even in experiments suitable to test for it.

(iii) Deformation lamellae 3–5 μm wide are formed in orientations of high resolved shear stress in some more anorthite-rich plagioclases (An_{37}, An_{44}, An_{77}) at 700–800°C and 10 kb. They may be due to slip along irrational planes lying 60–70° from [001] in unknown directions in these planes.

(iv) At any particular temperature and pressure, plagioclases of all compositions have very similar mechanical properties over the ranges

25–800°C and 5–10 kb confining pressure. As expected, the yield strength of polygrain plagioclase is one to three times greater than that of single crystals under the same conditions (Chapter 6).

(v) Deformation twinning cannot be produced in ordered albite, slip occurring instead, but if the albite is disordered by heating, twinning becomes dominant and slip is suppressed. Twinning has not been produced experimentally in peristerite either.

Natural Deformation of Plagioclase

Deformation twinning, on the albite and pericline laws, is common in natural plagioclase, even in grains that show no other optical evidence of strain. We must be wary of inferring too much about deformation conditions during metamorphism from this twinning, as it could have formed well after the main deformation(s), in response to local, even relatively weak stresses of relatively short duration.

A distinction between lamellar albite-law twins of growth and deformation origin can be made on the basis of shape, the former having planar interfaces and abrupt terminations, the latter commonly being lenticular.[55,56] Some lamellar twins produced experimentally by deformation have abrupt terminations,[9,10] but these are associated with optically visible terminal fractures, which appear to be absent from natural deformation twins.

Consideration of the atomic structure indicates that simple shear produces *pseudo-twins* in plagioclase, and that a considerable rearrangement (involving diffusion) of atoms must occur for a true twin to be produced.[49] So, natural deformation twins produced quickly and/or in relatively cold rocks, may in fact be pseudo-twins, and this may explain why recrystallisation and/or replacement by aggregates of secondary minerals occurs preferentially along certain twin lamellae in some rocks.[49] Some experimentally formed twins may also be pseudo-twins.[10]

Deformation bands have been observed in natural plagioclase (An_{40}). Some have the maximum misorientation across their boundaries of 7° found experimentally.[57] Bands observed in an albite (An_7) specimen have misorientations of up to 45°, but these have optically visible fractures along the band boundaries, and do not appear to be true kinks formed by plastic deformation.[48]

Experimental Recrystallisation of Plagioclase

I know of only one successful experiment on the recrystallisation of plagioclase.[34] A peristerite deformed at about 800°C and 10 kb in a

Griggs apparatus showed intense undulose extinction and 'mottling' in the central, most deformed part of the specimen. These optical features suggested that at least some plastic deformation had taken place, although an abundance of fractures also suggested that much of the deformation had been brittle. The transmission electron microscope revealed dislocations, partial dislocations (with associated stacking faults) and, in strongly deformed areas, abundant dislocation tangles. Also, in more deformed areas, local patches of polygonal new grains and subgrains occurred, indicating that recrystallisation had been achieved. However, systematic investigations still remain to be done.

Natural Recrystallisation of Plagioclase

Some spectacular examples of the apparent recrystallisation of deformed plagioclase have been described, but little systematic work has been done.

Partial recrystallisation of a large plagioclase (An_{40}) single crystal from a granulite facies terrain in Central Australia has occurred preferentially along narrow zones and in local patches of relatively high strain.[57] Deformation (kink) bands approximately or exactly normal to (001) are abundant, and have maximum misorientations across their boundaries of 7°, as in experimentally produced deformation bands. Local migration (bulging) of many band boundaries has occurred, the misorientation across these being up to 13°. In places, these bulges appear to have acted as nuclei for new (recrystallised) grains. A few deformation bands pass into diffuse patches of apparent subgrains that could give rise to eventual patches of recrystallised grains. Some possible fractures are marked by narrow zones of small, diffuse, equant subgrains that could give rise to recrystallised aggregates, so that sintering and growth of fragments cannot be excluded as one of the several possible nucleation mechanisms for the recrystallisation. Transmission electron microscopy is needed to help clarify the situation. Two points are clear, however, namely:

(i) The new (recrystallised) grains show crystallographic orientations close to that of the original grain, suggesting that a host/new grain relationship controls the intragranular recrystallisation of plagioclase, as it does in the intragranular recrystallisation of quartz and olivine (described earlier);

(ii) the new grains are slightly but consistently more albitic in composition than the original grain, as shown by electron microprobe analysis. More rocks need to be studied of course, but this may mean that for recrystallisation to occur in plagioclase there must be at least a small compositional change. This has been suggested also for biotite (see above), and possibly may apply to all complex silicate minerals. This is an interesting field of investigation, and may well illustrate the inter-dependence of chemical and physical processes in metamorphism. For example, the observed partial recrystallisation of plagioclase in some relatively low-grade rocks (referred to above) may have been induced more by chemical reactions going on in the rock, than by temperature. The idea is that chemical changes cause a reduction in free energy sufficient to surmount nucleation energy barriers that are too much for reduction in strain energy. In other words, for some reason (probably the complexity of atomic ordering in these compounds) dislocations cannot move easily, until chemical changes induce the necessary atom mobility and increased disordering of the structure.

In a transition from felsic gneiss to fine-grained mylonite in Central Australia, quartz shows internal deformation and partial recrystallisation at a stage at which feldspar shows only rare deformation twins and kink bands (with a few small new grains along some kink and grain boundaries).[6] However, as the rock becomes obviously foliated, the feldspar shows abundant deformation twinning and kinking, and subgrains and new grains are relatively common. New grains are most common in high strain areas, namely adjacent to grain boundaries or along zones of strong mechanical twinning or kinking. New grains are intimately associated with subgrains, suggesting that the nucleation mechanism in-volved (i) bulge nucleation along kink boundaries and (ii) subgrain growth and rotation during deformation. More recrystallisation relative to strain occurs in mylonites of the amphibolite facies (in which water is available) than in mylonites of the granulite facies in the same general mylonite zone.

Some remnants of large feldspar grains occur in the mylonite proper, although all original quartz and biotite grains have been completely recrystallised to fine-grained aggregates. This situation appears to be common in felsic mylonites, judging from published descriptions.

GENERALISATIONS

The foregoing discussion shows that, although much work has been car-ried out on the deformation, recovery and recrystallisation of common silicate minerals, a lot more remains to be done—both experimental studies and systematic observational work on natural rocks. However, some hypotheses seem reasonably tenable at the present stage of investigation (though they may need drastic alteration later on).

(i) Intragranular recrystallisation of silicates appears to involve strong crystallographic relationships between the original and the new grains. This is suggested by published work on quartz, olivine, biotite and plagioclase. Intergranular recrystallisation is more complicated or more difficult to study.

(ii) Intragranular recrystallisation of compositionally complex silicates (i.e. excluding quartz) appears to involve a change in chemical composition between new and old grains, as suggested by preliminary information on biotite and plagioclase. Whether we go as far as to suggest that recrystallisation in complex silicates will not occur at all unless accompanied by at least a small chemical change depends on how reckless we are. I would guess that this is so, but a great deal more work needs to be done. The idea is that stored plastic strain energy is not large enough to surmount the energy barriers to nucleation, although it appears to make a contribution, because recrystallisation generally occurs at sites of relatively high strain. Possibly chemical changes involving the breakage of strong bonds enable dislocations to move much more easily. Conceivably the chemical changes could involve (*a*) small, metastable compositional fluctuations ('spinodal decomposition') that act as nuclei in the original grains,[14, 16, 18, 22, 43] (*b*) stable exsolution or (*c*) larger chemical changes associated with reactions between various phases of the rock during metamorphism.

(iii) Deformation tends to occur by slip or twinning at lower temperatures and faster strain rates, but dynamic recovery and syntectonic recrystallisation take over at higher temperatures and slower strain rates. Naturally deformed minerals show evidence of all these processes, but transmission electron microscopy is needed as well as optical microscopy, in order to observe the density and arrangement of dislocations, before making firm interpretations.

(iv) Preferred orientations mainly develop as a result of intragranular slip (especially at lower temperatures and faster strain rates) or by recrystallisation (especially at higher temperatures and slower strain rates), or by a combination of both processes. Grain-boundary sliding may be important also, especially in materials with few slip systems (e.g. layer silicates), and mechanical rotation and growth constraints may be factors contributing to preferred orientations in these minerals.

References

1 Ave'Lallemant, H. G. (1967). Structural and petrofabric analysis of an 'alpine-type' peridotite: the lherzolite of the French Pyrenees. *Leidse Geologische Mededelingen*, **42**, 1–57.

2 Ave'Lallemant, H. G. and Carter, N. L. (1970). Syntectonic recrystallization of olivine and modes of flow in the upper mantle. *Bull. Geol. Soc. America*, **81**, 2203–20.

3 Ave'Lallemant, H. G. and Carter, N. L. (1971). Pressure dependence of quartz deformation lamellae orientations. *Amer. J. Science*, **270**, 218–35.

4 Bell, T. H. (1973). Mylonite development in the Woodroffe Thrust, north of Amata, Musgrave Ranges, central Australia. Unpub. Ph.D. thesis, Univ. Adelaide.

5 Bell, T. H. (1974). Development of quartz preferred orientation in mylonites of the Woodroffe Thrust, Central Australia, in press.

6 Bell, T. H. and Etheridge, M. A. (1973). Microstructure of mylonites and their descriptive terminology. *Lithos*, **6**, 337–48.

7 Blacic, J. D. and Christie, J. M. (1973). Dislocation substructure of experimentally deformed olivine. *Contribs. Mineralogy & Petrology*, **42**, 141–6.

8 Boland, J. N., McLaren, A. C. and Hobbs, B. E. (1971). Dislocations associated with optical features in naturally deformed olivine. *Contribs. Mineralogy & Petrology*, **30**, 53–63.

9 Borg, I. Y. and Heard, H. C. (1969). Mechanical twinning and slip in experimentally deformed plagioclases. *Contribs. Mineralogy & Petrology*, **23**, 128–35.

10 Borg, I. Y. and Heard, H. C. (1971). Experimental deformation of plagioclases, in *Experimental and Natural Rock Deformation*, ed. Paulitsch, P. 375–403. Berlin: Springer-Verlag.

11 Carter, N. L. (1971). Static deformation of silica and silicates. *J. Geophys. Research*, **76**, 5514–40.

12 Carter, N. L. and Ave'Lallemant, H. G. (1970). High temperature flow of dunite and peridotite. *Bull. Geol. Soc. Amer.*, **81**, 2181–202.

13 Carter, N. L., Christie, J. M. and Griggs, D. T. (1964). Experimental deformation and recrystallization of quartz. *J. Geol.*, **72**, 687–733.

14 Champness, P. E. and Lorimer, G. W. (1971). An electron microscopic study of a lunar pyroxene. *Contribs. Mineralogy & Petrology*, **33**, 171–83.

15 Christie, J. M., Griggs, D. T. and Carter, N. L. (1964). Experimental evidence of basal slip in quartz. *J. Geol.*, **72**, 734–56.

16 Christie, O. H. J. (1968). Spinodal precipitation in silicates. Introductory application to exsolution in feldspar. *Lithos*, **1**, 187–92.

17 Etheridge, M. A. (1971). Experimental investigations of the mechanisms of mica preferred orientation in foliated rocks. Unpub. Ph.D. thesis, Aust. National Univ.

18 Etheridge, M. A. and Hobbs, B. E. (1974). Chemical and deformational controls on recrystallization of mica. *Contribs. Mineralogy & Petrology*, **43**, 111–24.

19 Etheridge, M. A., Hobbs, B. E. and Paterson, M. S. (1973). Experimental deformation of single crystals of biotite. *Contribs. Mineralogy & Petrology*, **38**, 21–36.

20 Etheridge, M. A., Paterson, M. S. and Hobbs, B. E. (1973). Experimentally produced preferred orientation in synthetic mica aggregates. *Contribs. Mineralogy & Petrology*, **44**, 275–94.

21 Frondel, C. (1962). *Dana's System of Mineralogy*. Vol. III. *The Silica Minerals*. New York: J. Wiley & Sons, Inc.

22 Ghose, S., Phakey, P. P. and Tidy, E. (1972). Spinodal decomposition in an alkali amphibole. *Geol. Soc. America Abstracts*, **4**, 516.

23 Green, H. W. (1972). The nature of deformation lamellae in silicates. *Bull. Geol. Soc. America*, **83**, 847–52.

24 Green, H. W., Griggs, D. T. and Christie, J. M. (1970). Syntectonic and annealing recrystallization of fine-grained quartz aggregates, in *Experimental and Natural Rock Deformation*, ed. Paulitsch, P. 272–335. Berlin-Heidelberg-New York: Springer-Verlag.

25 Green, H. W. and Radcliffe, S. V. (1972). Deformation processes in the upper mantle, in *Flow and Fracture of Rocks, Amer. Geophysical Union, Monograph*, ed. Heard, H. C. *et al.*, **16**, 139–56.

26 Griggs, D. T. (1967). Hydrolytic weakening of quartz and other silicates. *Geophys. J. Roy. Astron. Soc.*, **14**, 19–32.

27 Griggs, D. T. and Blacic, J. D. (1965). Quartz: anomalous weakness of synthetic crystals. *Science*, **147**, 292–5.

28 Hobbs, B. E. (1968). Recrystallization of single crystals of quartz. *Tectonophysics*, **6**, 353–401.

29 Kehlenbeck, M. M. (1972). Deformation textures in the Lac Rouvray anorthosite mass. *Canadian J. Earth Sci.*, **9**, 1087–98.

30 McLaren, A. C. and Hobbs, B. E. (1972). Transmission electron microscope investigation of some naturally deformed quartzites, in *Flow and Fracture of Rocks. Amer. Geophysical Union, Monograph*, ed. Heard, H. C. *et al.*, **16**, 55–66.

31 McLaren, A. C. and Retchford, J. A. (1969). Transmission electron microscope study of the dislocations in plastically deformed synthetic quartz. *Physica Status Solidi*, **33**, 657–68.

32 McLaren, A. C., Retchford, J. A., Griggs, D. T. and Christie, J. M. (1967). Transmission electron microscope study of Brazil twins and dislocations experimentally produced in natural quartz. *Physica Status Solidi*, **19**, 631–44.

33 McLaren, A. C., Turner, R. G., Boland, J. N. and Hobbs, B. E. (1970). Dislocation structure of the deformation lamellae in synthetic quartz; a study by electron and optical microscopy. *Contribs. Mineralogy & Petrology*, **29**, 104–15.

34 Marshall, D., Hobbs, B. E. and Vernon, R. H. Experimental deformation and recrystallization of a peristerite (in preparation).

35 Maxwell, J. C. (1962). Origin of slaty and fracture cleavage in the Delaware Water Gap Area, New Jersey and Pennsylvania, in *Petrologic studies. Amer. J. Science*, ed. Engel, A. E. J., James, H. L. and Leonard, B. E. 281–311.

36 Means, W. D. and Paterson, M. S. (1966). Experiments on preferred orientation of platy minerals. *Contribs. Mineralogy & Petrology*, **13**, 108–33.

37 Means, W. D. and Williams, P. F. (1972). Crenulation cleavage and faulting in an artificial salt-mica schist. *J. Geology*, **80**, 569–91.

38 Möckel, J. R. (1969). Structural petrology of the garnet-peridotite of Alpe

Arami (Ticino, Switzerland). *Leidse Geologische Mededelingen*, **42**, 61–130.

39 Moore, A. C. (1973). Studies of igneous and tectonic textures and layering in the rocks of the Gosse Pile intrusion, Central Australia. *J. Petrology*, **14**, 49–80.

40 Nicolas, A., Bouchez, J. L., Boudier, F. and Mercier, J. C. (1971). Textures, structures and fabrics due to solid state flow in some European lherzolites. *Tectonophysics*, **12**, 55–86.

41 Nicolas, A., Boudier, F. and Boullier, A. M. (1973). Mechanisms of flow in naturally and experimentally deformed peridotites. *Amer. J. Science*, **273**, 853–76.

42 Olsen, A. and Birkeland, T. (1973). Electron microscope study of peridotite xenoliths in kimberlites. *Contribs. Mineralogy and Petrology*, **42**, 147–57.

43 Owen, D. C. and McConnell, J. D. C. (1971). Spinodal behaviour in an alkali feldspar. *Nature Physical Science*, **230**, 118–19.

44 Phakey, P., Dollinger, G. and Christie, J. M. (1972). Transmission electron microscopy of experimentally deformed olivine crystals, in *Flow and Fracture of Rocks*. *Amer. Geophysical Union, Monograph*, ed. Heard, H. C. *et al.*, **16**, 117–38.

45 Powell, C. McA. (1969). Intrusive sandstone dykes in the Siamo Slate near Negaunee, Michigan. *Bull. Geol. Soc. America*, **80**, 2585–94.

46 Raleigh, C. B. (1968). Mechanisms of plastic deformation of olivine. *J. Geophys. Research*, **73**, 5391–406.

47 Ransom, D. M. (1971). Host control of recrystallized quartz grains. *Mineralogical Mag.*, **38**, 83–8.

48 Seiffert, K. E. (1965). Deformation bands in albite. *Amer. Mineralogist*, **50**, 1469–72.

49 Starkey, J. (1964). Glide twinning in the plagioclase feldspars, in *Deformation Twinning*, ed. Reed-Hill, R. E. *et al.* 177–91. New York: Gordon & Breach.

50 Tullis, J. A. (1968). Preferred orientation in experimental quartz mylonites. *Trans. Amer. Geophys. Union*, **39**, 755.

51 Tullis, J. A. (1970). Quartz: preferred orientation in rocks produced by Dauphiné twinning. *Science*, **168**, 1342–4.

52 Tullis, J. A. (1971). Preferred orientations of experimentally deformed quartzites. Unpub. Ph.D. thesis, Univ. California, Los Angeles.

53 Tullis, J. A., Christie. J. M. and Griggs, D. T. (1973). Microstructures and preferred orientations of experimentally deformed quartzites. *Bull. Geol. Soc. America*, **84**, 297–314.

54 Tullis, T. E. (1971). Experimental development of preferred orientation of mica during recrystallization. Unpub. Ph.D. thesis, Univ. California, Los Angeles.

55 Vance, J. A. (1961). Polysynthetic twinning in plagioclase. *Amer. Mineralogist*, **46**, 1097–119.

56 Vernon, R. H. (1965). Plagioclase twins in some mafic gneisses from Broken Hill, Australia. *Mineralogical Mag.*, **35**, 488–507.

57 Vernon, R. H. (1975). Natural intragranular recrystallization of plagioclase (in preparation).

58 White, S. (1971). Natural creep deformation of quartzites. *Nature Physical Science*, **234**, 175–7.

59 White, S., Crosby, A. and Evans, P. E. (1971). Dislocations in naturally deformed quartz. *Nature Physical Science*, **231**, 85–6.
60 White, S. (1973). The dislocation structures responsible for the optical effects in some naturally-deformed quartzes. *J. Materials Science*, **8**, 490–9.
61 Wilson, C. J. L. (1973). The prograde microfabric in a deformed quartzite sequence, Mount Isa, Australia. *Tectonophysics*, **19**, 39–81.
62 Wilson, C. J. L. and Glass, J. (1974). Preferred orientation in quartz ribbon mylonite (in press).
63 Yar Khan, M. (1972). The structure and microfabric of a part of the Arltunga Nappe Complex, central Australia. Unpub. Ph.D. thesis, Australian National Univ., Canberra.

Chapter 8

Relationships between Chemical and Physical Processes in Metamorphism

Introduction

Most metamorphic investigations have been concerned with either chemical or physical aspects, but not both. However, a few people have tried to integrate the two, and the problem is getting increasing attention. This approach appears fundamental to the understanding of metamorphic processes. The effects of non-hydrostatic stress on the equilibrium conditions of a coherent phase transformation ($\alpha \rightleftharpoons \beta$ quartz) have been mentioned in Chapter 2, and the possible effects of deformation on metamorphic reaction kinetics have been mentioned in Chapters 3 and 4. These are among the most important relationships between chemical and physical aspects of metamorphism, and should always be kept in mind when thinking about metamorphic processes. The basis for the application of non-hydrostatic thermodynamics in geological processes has been reviewed by Paterson.[24]

In this chapter I want to mention (i) the possible effect of chemical reactions on the mechanical properties of the rocks concerned and, conversely, (ii) the effect of deformation on the mineral and/or chemical composition of the rock. These topics are worthy of much more investigation, both in natural rocks and experimental situations.

Effect of Chemical Reactions on Rock Mechanical Properties

Experimental deformation of natural serpentinite has shown that, above the temperature at which serpentine dehydrates to talc and forsterite (Fig. 4.11), the serpentinite becomes embrittled and deforms readily along certain thin zones in the specimen.[26] This is an example of the possible effect of chemical reactions on mechanical properties of metamorphic rocks.

Holland and Lambert have suggested a general scheme relating the rheology of rocks to metamorphic facies.[16] Although relationships are idealised, the approach is interesting. They noted that one rock unit might be capable of undergoing metamorphic reaction with associated alteration of its creep-rate, whereas a neighbouring unit could be metamorphically stable and show different creep properties. Their generalised scheme is as follows:

Regime 1: pre-metamorphic or very low-grade conditions and elastic deformation;

Regime 2: low-grade metamorphic conditions (especially involving the production of micas and amphiboles), and flow producing a wide variety of structures; strain rates are high but depend on the chemical composition of the rock;

Regime 3: upper greenschist facies and most of the amphibolite facies, with deformation by creep at lower rates than in Regime 2, and with similar rheologies for all common rock-types;

Regime 4: upper amphibolite facies conditions, in which dehydration reactions and partial melting cause an increase in the creep-rate, producing large strains that vary greatly with the rock-type;

Regime 5: granulite facies and upper mantle conditions, with deformation by prolonged laminar flow leading ultimately to simple styles of deformation, all rock-types having similar rheology.

A more specific example of the relationship between deformation and (apparent) metamorphic grade is Bell's study of the Woodroffe mylonite zone in Central Australia.[3] He found that availability of water on one side of the zone produced (i) amphibolite facies assemblages, as opposed to 'dry', granulite facies assemblages on the other side, and (ii) different mechanisms and degrees of recrystallisation on each side (see Chapter 7).

Effect of Differential Deformation on Mineral Assemblage

The reverse situation is possible also, namely the ability of deformation to control the mineral assemblage, by causing migration of components to various sites in a heterogeneously deforming rock.

Many metamorphic rocks (especially those formed in regional terrains) have continuous or discontinuous compositional foliations and/or lineations. For example, Talbot and Hobbs described compositional layering of metamorphic origin cutting recognisable relic bedding in metapelitic rocks of very low to moderate metamorphic grades.[30] In some of these rocks, the layering is parallel to slaty cleavage or crenulation cleavage, suggesting a structural control, but in other rocks the layering is independent of planar tectonic structures. Mostly the layering occurs in rocks showing evidence of heterogeneous strain, but in some rocks a strong layering occurs in domains of no apparent differential strain.

Ramberg suggested that differential *stress* during metamorphism could cause segregation of more mobile elements (Si, Al, K and Na, according to his scheme) into low-pressure zones (e.g. tension cracks), causing a 'metamorphic differentiation' (i.e. production of compositional layering).[27] The segregation occurs in response to *chemical potential gradients*, the behaviour of which has been mentioned in Chapter 4, in relation to compositional exchange in the absence of deformation. Others have applied this model, especially to layered mafic rocks, suggesting that the foliation planes acted as loci of low pressure, although how this could come about is not clear.[5, 10] Presumably we are to regard the thermodynamic pressure in a deforming rock as being approximately equal to the mean stress ($\bar{\sigma} = \sigma_1 + \sigma_2 + \sigma_3)/3$,[8] so that the 'pressure gradients' referred to are actually $\bar{\sigma}$ gradients. Carpenter[7] suggested that constituents with the largest partial molar volumes should experience the greatest changes in chemical potential and should be the first to migrate in response to a pressure gradient. In this way, he explained why relatively large ions and molecules, such as K^+, Na^+, H_2O and CO_2, seem to be among the most mobile components in metamorphic rocks.

An important suggestion of Bennington is that dense Mg- and Fe-rich mineral assemblages may form as residual accumulations in regions of low pressure in a heterogeneously deforming rock, these regions undergoing a volume decrease owing to depletion of less dense material (e.g. SiO_2).[4] He stated that the new mineral assemblage contains minerals that can grow from the remaining material under the existing stress, temperature and pressure conditions. This provides a possible theoretical basis for interpretations of the origin of compositional layering related to crenulation foliation, whereby felsic material is preferentially removed from (and micaceous material preferentially concentrated in) high-pressure zones in the appressed limbs of the asymmetrical crenulations.[15, 20, 21, 25, 33] A similar model involves concentration of felsic material in fold hinges during isoclinal folding, followed by increased appression to produce a gneissic layering.[1]

Not only differences in minerals or mineral assemblages, but also differences in the composition of individual minerals, have been ascribed to differential stress. For example, detailed electron microprobe analyses have been carried out on amphiboles occurring at the inner and outer arcs in the hinge zone of a folded amphibolite layer about 1 cm thick.[29] Amphibole grains in the inner arcs of the folded layer are Mg-rich, whereas those in the outer arcs are Fe-rich. Stress analysis of the folding of a competent layer in an incompetent matrix indicates a low-pressure area along the outer arc of the competent layer and a high-pressure area along the inner arc.[29] Also, the highest pressure gradient is directed along the axial plane of the fold. Substitution of Mg^{2+} (0·66Å) in preference to Fe^{2+} (0·74Å) would produce a denser Mg-amphibole in the high-pressure arc in the core of the fold, as is observed.

Differential *strain*, rather than local stress differences, may be an important control in some metamorphic differentiation. For instance, quartz aggregates in some mylonitic rocks are arranged in microfabric domains (defined by their shape and the preferred orientation of [001] of the quartz grains within them) that have gradational boundaries and are distributed regularly in relation to folds.[14] An idealised model has been used to explain the fabric domains, on the basis of different strain rates between adjacent shear domains.[14] In this example, a compositional heterogeneity is not produced, but it seems reasonable to keep the process in mind when considering the development of compositional heterogeneities.

Examples of Metamorphic Layering Formed by Preferential Removal of Minerals

Using partly dissolved fossils, Plessman has shown that mica-rich films in some carbonate-rich rocks were formed by preferential solution and removal of non-micaceous material.[25] Similarly, Williams showed that selective solution and removal of quartz produced micaceous layers in low-grade metasedimentary rocks at Bermagui, Australia.[33] Mica-poor layers in these rocks show detrital microstructures, whereas mica-rich layers show strong preferred orientations formed by mechanical rotation of residual detrital mica grains. In both these examples, a decrease in volume of former rock now occupied by mica-rich material has been demonstrated. At Bermagui, migration of quartz is not restricted to lateral movement from micaceous to adjacent quartz-rich domains, but must have extended to distances at least as great as half the wavelength of mesoscopic folds showing compositional differences between limbs and closures. Williams proposed that variation of $\bar{\sigma}$ in different parts of folds can lead to preferential solution of quartz in more highly stressed areas.[33]

In a detailed study of deformation and metamorphism near Ducktown, Tennessee, Holcombe inferred compositional segregation of micaceous versus felsic (and/or carbonate-rich) material, in slaty cleavage, crenulation cleavage, 'pressure shadows' and possibly quartz and carbonate veins occurring mostly along axial planes of folds.[15] Since the minerals in apparently precipitation situations, such as pressure shadows and veins, are mainly quartz (less commonly carbonate and feldspar), Holcombe and others have suggested that the components of these minerals are mobile, whereas mica and fine-grained TiO_2-rich materials are immobile and therefore are concentrated as a residuum in this segregation process. Probably the mineral segregation occurred without additional material from a distant external source, since the combined chemical composition of both micaceous and felsic (or calcareous) material approximates a reasonable sedimentary composition. However, on the scale of local compositional layering, the mobile material appears to have moved out of the local system, to be deposited nearby as veins or pressure shadows.*

The driving force for the segregation could be a combination of (*a*) a difference between volume change components of strain between different domains of a heterogeneously deforming rock, and (*b*) a chemical potential gradient caused by a stress difference between domains (mentioned above). Holcombe favoured (*a*) as the largest contributor, especially in the formation of slaty and crenulation cleavages, whereas he suggested that (*b*) would predominate where volume change differences are insignificant (possibly in the development of some 'pressure shadows').[15] The reason volume change can lead to concentration or depletion of different minerals is that the transport kinetics may differ for each mineral.[15] Another reasonable suggestion is that minerals of large specific volume may be unstable in domains undergoing volume decrease, and so would be selectively removed (as discussed above).[4]

However, not all compositionally layered deformed rocks need to have been formed by a metamorphic differentiation process. For example, compositional layering associated with crenulation folding has been produced experimentally under conditions that appear to have involved only mechanical deformation.[20] Moreover, simple deformation and recrystallisation of coarse-grained polyphase igneous or metamorphic rocks in mylonitic zones can produce compositional foliation and/or layering.[31] This is simply because the grains in the original rock were much larger than those in the resulting mylonite, so that the new distorted, recrystallised aggregates automatically have different compositions because each one represents a different parent grain (Fig. 8.1). Of course, if chemical

*Vidale (1974, *Bull. Geol. Soc. America*, **85**, 303–6) has observed that vein and pressure-shadow mineral assemblages vary consistently with metamorphic grade over an area of 1 700 km^2 in metapelitic rocks in and near Dutchess County, New York. Microprobe analyses show that the plagioclase of veins in the staurolite zone has a similar composition ($\pm An_5$) to the plagioclase in the adjacent rock. It appears that the vein material was derived locally from the surrounding matrix, possibly in response to $\bar{\sigma}$ gradients. More detailed studies linking metamorphism, deformation and vein asemblages are needed.

reactions occur during mylonitisation, this suggestion no longer applies; in fact, compositional layering may be reduced, rather than intensified,

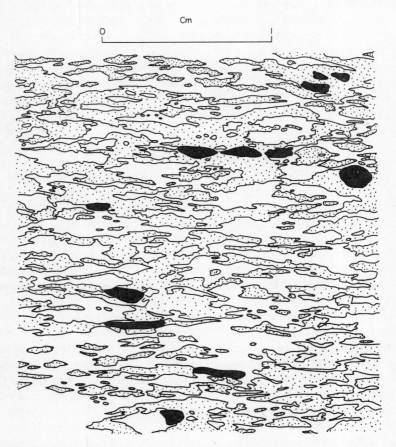

Fig. 8.1   Mafic blastomylonite from the Giles Complex, Central Australia, showing a lenticular compositional layering consisting of elongate recrystallised aggregates of plagioclase (clear) and pyroxene (stippled). Large, deformed pyroxene relics are shown in black. Each aggregate is interpreted as representing the distorted shape of a former large plagioclase or pyroxene grain in the original gabbroic rock.

by the tendency of new minerals to nucleate dispersely through the rock, so that grains of like phases become separated as much as possible (as discussed in Chapter 5).[3]

Large-Scale Metasomatism and Deformation

In some areas, the transfer of material in response to differential deformation appears to have occurred on a relatively large scale. An example is the production of foliated hydrous, oxidised, relatively K-rich rocks in shear zones in formerly less oxidised, anhydrous, K-poor rocks north of Scourie in Scotland.[2] It appears that metasomatism involving the introduction of potassium, water and oxygen, and the removal of silica, took place in zones of relatively more intense deformation.

Carpenter has suggested that folding during a metamorphic episode can bring about local concentration of water in fold hinges and thereby cause the development of hydrous assemblages that conventionally would be regarded as retrograde.[7] He called this 'apparent retrograde metamorphism', because it is inferred to have occurred during the main metamorphism. As example, he cited a folded metabasalt flow in the Moppin metavolcanic sequence of New Mexico, USA, in which this apparent retrograde metamorphism is expressed as a hydrous chlorite-albite-bearing assemblage in large fold hinges, resulting from the destruction late in the regional metamorphism of the hornblende-, oligoclase- and biotite-bearing assemblages found elsewhere in this unit.

Pegmatites and Deformation

Following Ramberg's[27] general suggestion that pegmatites are localised in low-pressure areas in metamorphic rocks, Gresens[11, 12] has proposed a model for the formation of some zoned pegmatites in schist terrains, involving diffusion of alkali chloride hydrothermal pore solution (in equilibrium with the solid phases) into zones of relatively low pressure. The fluid could be of either metamorphic or magmatic origin, and the presence of the fluid is postulated in view of the common occurrence of alkali chloride solutions as fluid inclusions in minerals (Chapter 2). Experimentally determined phase equilibria indicate that, with decreasing pressure, feldspar can form from muscovite, quartz and alkalis in the fluid phase. This could account for the development of quartz-feldspar pegmatites in low-pressure zones, whatever their origin. Once the pressure is raised again to normal lithostatic pressure, the feldspar could revert to muscovite, which is a common late-stage process in these pegmatites.

Migmatites

Much controversy surrounds the problem of the origin of migmatites. Experimental work has given information on the composition of melts to be expected by partial melting of various bulk compositions at various temperatures. The felsic components of some migmatites accord with the composition of minimum melts to be expected, but in other migmatites the felsic components do not.[32] Furthermore, in some rocks of mafic

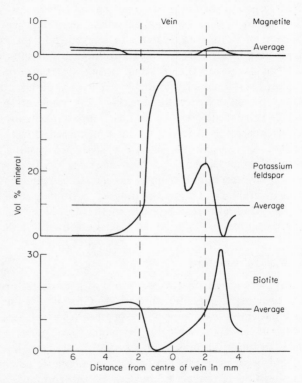

Fig. 8.2   Modal distribution of magnetite, K-feldspar and biotite in a granitic vein and adjacent host rock, Palmer migmatite area, South Australia. Note the relative enrichment of biotite and magnetite in the vein selvedges. Horizontal lines give the average modal composition of the bulk rock. After White, p. 179.[32]

composition, quartzofeldspathic layers contain plagioclase of the same composition as that of the host-rock, eliminating an origin by partial melting.[19] Mafic selvedges and quartzofeldspathic veins (or layers) commonly have complementary compositions (Figs. 8.2 and 8.3),[13, 19, 32] which suggests local lateral segregation of components down chemical potential gradients, but which need not exclude an origin by local partial melting. Hughes suggested that the selvedges originate at boundaries between rock with a hydrothermal pore-fluid and rock with a silicate pore-fluid formed during the incipient stages of partial melting.[17] Some detailed studies involving minor elements (especially Rb, Ba, and Sr) favour

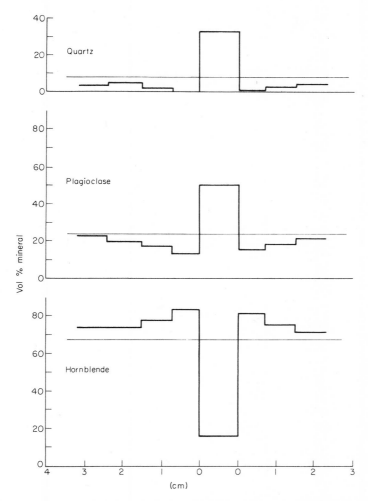

Fig. 8.3 Histogram profiles of modal variation (in volume per cent) with distance from a quartz-feldspar vein in amphibolite, Einasleigh, North Queensland. Horizontal lines represent composition of the bulk rock. After Kretz, p. 567.[19]

lateral segregation, rather than partial melting.[13,32] Whatever the origin of migmatites, a relationship between chemical change and deformation appears unlikely, at least in many examples, because the migmatisation commonly appears to have post-dated recognisable deformation effects.

Time Relationships Between Deformation and
Mineral Growth[18, 28, 34, 35, 36]

Inferences made from microscopic examination of the arrangement of inclusions in porphyroblasts and the arrangement of the foliation in the surrounding matrix have resulted in a well established technique ('tex-

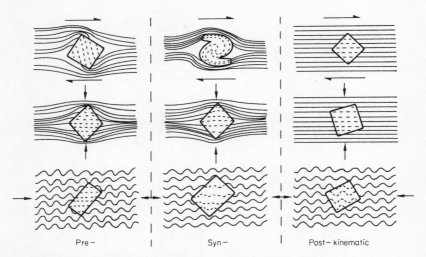

Pre−   Syn−   Post−kinematic

Fig. 8.4   Diagrammatic representation of nine forms of porphyroblasts believed to be diagnostic with regard to time of growth of porphyroblast and time of development of foliation. After Zwart, p. 41.[36]

tural analysis') for relating time, deformation (production of foliation) and mineral growth. The main criteria are summarised in figure 8.4, and an example of their application to a particular metamorphic area is given in figure 8.5. The technique suggests that in some areas several metamorphic 'pulses' may have occurred, although in others a broad metamorphic event appears to have encompassed several 'episodes' of deformation. The approach is based on 'commonsense' interpretation of microstructures, rather than experimental results, but it has considerable value, provided mineral assemblages, rather than individual minerals, are considered. If this is not done, episodes of growth of individual minerals may be postulated (as has been done at times), which, because most minerals have chemical compositions different from that of the bulk rock, implies a metasomatic episode for each mineral.[6] Moreover, as discussed in Chapter 2, some inferred metamorphic reactions involve

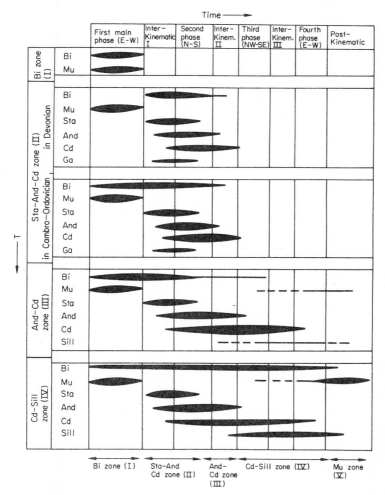

Fig. 8.5 Tabulated relationships between time, metamorphic zones and deformation phases inferred by 'textural analysis' of rocks of the Bosost area (central Pyrenees). After Zwart, p. 48.[36]

the formation of a mineral and its inclusions (e.g. biotite with sillimanite) as products of the same reaction, which adds another complication to 'textural analysis' of metamorphic rocks.[6]

Microboudinage of zoned grains and of grains showing inferred partial reaction to other minerals has been used ingeniously by Misch as an indicator of paracrystalline deformation (i.e. synkinematic growth), as shown in figures 8.6 and 8.7.[22, 23]

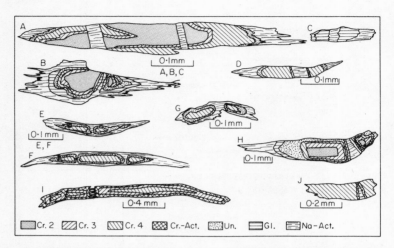

Fig. 8.6 Paracrystalline microboudinage of grains (now optically continuous relics) of Na-amphibole in fine-grained 'blueschists' in the northern Cascades, Washington, USA. Cr=crossite; Cr-Act=crossactinolite; Un='uniaxial' blue amphibole; Gl=glaucophane; Na-Act=sodic actinolite. After Misch, p. 46.[22]

However, each rock and area must be treated individually, and, if a sequence of deformation and mineral growth can be worked out for part of an area, it need not necessarily apply to another part of the same area. This is because the type of strain, amount of strain, strain rate, heating rate and volatile content may all vary independently in time and place.

Solution Transfer

Many people have suggested that crystalline material may be dissolved at sites of relatively high stress and precipitated at sites of lower stress. This is thermodynamically feasible,[8, 27] especially where grains are in partial contact and a pore fluid is present, but we do not know for certain that the magnitude of the effect is large enough to produce appreciable compositional change in all metamorphic situations. Pressure solution offers a good explanation of stylolytic surfaces, some of which are demonstrably tectonic in origin, in limestones and quartzites. It may also assist deformation under certain conditions,[8] but it is not necessarily a mechanism for causing chemical change. However, it may do so, especially if it is really the way in which 'pressure shadows' and related microstructures are produced (as is commonly suggested[8, 28]), because these typically are mineral concentrations.

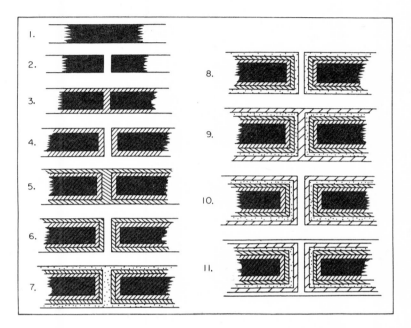

Fig. 8.7 Diagrammatic deformation-growth sequence involved in paracrystalline microboudinage of zoned grains, showing a single site of rupture. Although represented graphically as finite steps, both stretching and accretion are thought to proceed essentially simultaneously, the steps being infinitesimal. After Misch, p. 48.[22]

Spinodal Decomposition?

Etheridge and Hobbs have suggested that small compositional fluctuations may give rise to nuclei of new grains in deformed biotite.[9] This 'spinodal decomposition' may be more favourable kinetically than conventional nucleation, under some circumstances, although it is thermodynamically unfavourable. If it operates, either a metastable phase or a metastable composition of the original phase may grow, which possibility should be kept in mind when considering compositional relationships between phases in metamorphic rocks. However, the extent of the applicability of the process to metamorphic rocks (if at all) is unknown.

References

1 Ayrton, S. (1969). On the origin of gneissic banding. *Eclogae Geol. Helv.*, **62**, 567–70.

2 Beach, A. and Fyfe, W. S. (1972). Fluid transport and shear zones at Scourie, Sutherland: evidence of overthrusting? *Contribs. Mineralogy & Petrology*, **36**, 175–80.

3 Bell, T. H. (1973). Mylonite development in the Woodroffe Thrust, north of Amata, Musgrave Ranges, central Australia. Unpub. Ph.D. thesis, Univ. Adelaide.

4 Bennington, K. O. (1956). Role of shearing stress and pressure in differentiation as illustrated by some mineral reactions in the system $MgO-SiO_2-H_2O$. *J. Geology*, **64**, 558–77.

5 Bowes, D. R. and Park, R. G. (1966). Metamorphic segregation banding in the Loch Kerry basite sheet from the Lewisian of Gairloch, Ross-shire, Scotland. *J. Petrology*, **7**, 306–30.

6 Carmichael, D. M. (1969). On the mechanism of prograde metamorphic reactions in quartz-bearing pelitic rocks. *Contribs. Mineralogy & Petrology*, **20**, 244–67.

7 Carpenter, J. R. (1968). Apparent retrograde metamorphism: another example of the influence of structural deformation on metamorphic differentiation. *Contribs. Mineralogy & Petrology*, **17**, 173–86.

8 Durney, D. W. (1972). Solution transfer, an important geological deformation mechanism. *Nature*, **235**, 315–17.

9 Etheridge, M. A. and Hobbs, B. E. (1974). Chemical and deformational controls on recrystallization of mica. *Contribs. Mineralogy & Petrology*, **43**, 111–24.

10 Ghaly, T. S. (1969). Metamorphic differentiation in some Lewisian rocks of north west Scotland. *Contribs. Mineralogy & Petrology*, **22**, 276–89.

11 Gresens, R. L. (1967). Tectonic-hydrothermal pegmatites. I. The model. *Contribs. Mineralogy & Petrology*, **15**, 345–55.

12 Gresens, R. L. (1967). Tectonic-hydrothermal pegmatites. II. An example. *Contribs. Mineralogy & Petrology*, **16**, 1–28.

13 Hedge, C. E. (1972). Source of leucosomes of migmatites in the Front Range, Colorado. *Geol. Soc. Amer. Mem.*, **135**, 65–72.

14 Hobbs, B. E. (1966). Microfabric of tectonites from the Wyangala Dam area, New South Wales, Australia. *Bull. Geol. Soc. America*, **77**, 685–706.

15 Holcombe, R. J. (1973). Mesoscopic and microscopic analysis of deformation and metamorphism near Ducktown, Tennessee. Unpub. Ph.D. thesis, Stanford University.

16 Holland, J. G. and Lambert, R. St. J. (1969). Structural regimes and metamorphic facies. *Tectonophysics*, **7**, 197–217.

17 Hughes, C. J. (1970). The significance of biotite selvedges in migmatites. *Geol. Mag.*, **107**, 21–4.

18 Johnson, M. R. W. (1963). Some time relations of movement and metamorphism in the Scottish Highlands. *Geologie en Mijnbouw*, **42**, 121–42.

19 Kretz, R. (1966). Metamorphic differentiation at Einasleigh, north Queensland. *J. Geol. Soc. Australia*, **13**, 561–82.

20 Means, W. D. and Williams, P. F. (1972). Crenulation cleavage and faulting in an artificial salt-mica schist. *J. Geology*, **80**, 569–91.

21 Means, W. D. and Williams, P. F. (1974). Compositional differentiation in an experimentally deformed salt-mica specimen. *Geology*, **2**, 15–16.
22 Misch, P. (1969). Paracrystalline microboudinage of zoned grains and other criteria for synkinematic growth of metamorphic minerals. *Amer. J. Science*, **267**, 43–63.
23 Misch, P. (1970). Paracrystalline microboudinage in a metamorphic reaction sequence. *Bull. Geol. Soc. America*, **81**, 2483–6.
24 Paterson, M. S. (1973). Nonhydrostatic thermodynamics and its geologic applications. *Reviews of Geophysics & Space Physics*, **11**, 355–89.
25 Plessman, W. (1964). Gesteinslösun, ein Hauptfaktor beim Schieferungsprozess. *Geol. Mitt.*, **4**, 69–82.
26 Raleigh, C. B. and Paterson, M. S. (1965). Experimental deformation of serpentinite and its tectonic implications. *J. Geophysical Research*, **70**, 3965–85.
27 Ramberg, H. (1952). *The Origin of Metamorphic and Metasomatic Rocks.* Chicago: Univ. Chicago Press.
28 Spry, A. (1969). *Metamorphic Textures.* Oxford: Pergamon Press.
29 Stephansson, O. (1974). Stress-induced diffusion during folding. *Tectonophysics*, **22**, 233–51.
30 Talbot, J. L. and Hobbs, B. E. (1968). The relationship of metamorphic differentiation to other structural features at three localities. *J. Geology*, **76**, 581–7.
31 Vernon, R. H. (1974). Controls of mylonitic compositional layering during non-cataclastic ductile deformation. *Geol. Mag.*, **111**, 121–3.
32 White, A. J. R. (1966). Genesis of migmatites from the Palmer region of South Australia. *Chemical Geology*, **1**, 165–200.
33 Williams, P. F. (1972). Development of metamorphic layering and cleavage in low grade metamorphic rocks at Bermagui, Australia. *Amer. J. Science*, **272**, 1–47.
34 Wilson, M. R. (1971). On syntectonic porphyroblast growth. *Tectonophysics*, **11**, 239–60.
35 Zwart, H. J. (1960). The chronological succession of folding and metamorphism in the central Pyrenees. *Geologischen Rundschau*, **50**, 203–18.
36 Zwart, H. J. (1962). On the determination of polymetamorphic mineral associations, and its application to the Bosost area (central Pyrenees). *Geologischen Rundschau*, **52**, 38–65.

Subject Index

Author Index

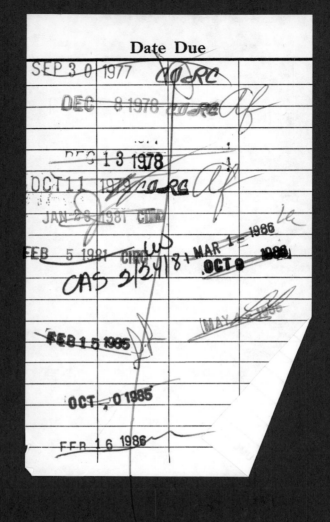